ANNUAL REPORT

Center for Experimental Nuclear Physics and Astrophysics
University of Washington
August 29, 2018

Sponsored in part by the United States Department of Energy
under Grant #DE-FG02-97ER41020.

This report was prepared as an account of work sponsored in part by the United States Government. Neither the United States nor the United States Department of Energy, nor any of their employees, makes any warranty, expressed or implied or assumes any legal liability or responsibility for accuracy, completeness or usefulness of any information, apparatus, product or process disclosed, or represents that its use would not infringe on privately-owned rights.

Cover design by Gary Holman. Photos from left: Nathan Froemming inspecting upstream muon beam line for the Muon $g - 2$ experiment. Krishna Venkateswara assembling one of the four tiltmeters at LIGO Livingston Observatory. Brent Graner and Drew Byron examining the superconducting solenoid to be used for the ^{6}He-CRES experiment. David Hertzog calibrating the beam-entrance counter for the Muon $g - 2$ experiment. Ali Ashtari Esfahani while exchanging a quarter-wave plate assembly on the Project 8 Phase II tritium compatible waveguide setup.

INTRODUCTION

The Center for Experimental Nuclear Physics and Astrophysics, CENPA, was established in 1998 at the University of Washington as the institutional home for a broad program of research in nuclear physics and related fields. Research activities — with an emphasis on fundamental symmetries and neutrinos — are conducted locally and at remote sites. In neutrino physics, CENPA is the lead US institution in the KATRIN tritium β-decay experiment, the site for experimental work on Project 8, and a collaborating institution in the MAJORANA ^{76}Ge $0\nu\beta\beta$ experiment. The Muon Physics group has developed the MuSun experiment to measure muon capture in deuterium at the Paul Scherrer Institute in Switzerland. The group has a leadership role in the new project to measure the anomalous magnetic moment of the muon at Fermilab to even higher precision than it is presently known from the collaboration's previous work at Brookhaven. The fundamental symmetries program also includes "in-house" research on the search for a static electric dipole moment in ^{199}Hg, and an experiment using the local tandem Van de Graaff accelerator to measure the electron-neutrino correlation and Fierz interference in ^{6}He decay.

In addition to the research directly supported by DOE's Office of Nuclear Physics through the CENPA core grant, other important programs are located at CENPA, forming a broader intellectual center with valuable synergies. The "Gravity" group, as it is known, carries out with both DOE and NSF support studies of the weak and strong Equivalence Principles, as well as searches for non-Newtonian forces such as those predicted by theories with extra dimensions. In addition, they participate in LIGO. The DOE Office of High Energy Physics supports the unique ADMX axion search experiment.

CENPA is home to a large number of faculty, research faculty, postdoctoral researchers, graduate, and undergraduate students. The core professional engineering and technical staff provide diverse capabilities and skills such as state-of-the-art detector development, fabrication of custom electronics, large-scale computing, and design engineering. New advancements, capabilities, and ideas are regularly shared at seminars by CENPA members and visitors alike.

Victoria Clarkson

This year we were all saddened by the death of Victoria Clarkson after her courageous battle with cancer. She was our highly regarded and loved administrator and will be deeply missed.

Notable

The DOE Office of Nuclear Physics has awarded CENPA a new three-year continuation of the program, which is supporting the bulk of the efforts described in this Annual Report.

Transitions

Alvaro Chavarria joined the University of Washington faculty in January, 2018 as a new Assistant Professor and member of the CENPA faculty. Prof. Chavarria is the Analysis Coordinator of the DAMIC dark matter experiment and he is the conceptual lead for the new Selena neutrinoless $0\nu\beta\beta$ experiment. He has begun his laboratory work in a repurposed CENPA clean room located in the main Physics building. Prof. Jason Detwiler was promoted to Associate Professor of Physics with tenure, starting in Fall 2018. Martin Fertl and Jarek Kaspar were both appointed to be Research Assistant Professors.

Peter Doe was named U.W. Spokesperson for KATRIN, succeeding Hamish Robertson. Jason Detwiler was named Co-Spokesperson for the MAJORANA DEMONSTRATOR.

Postdoc Brent Graner, who completed his thesis on the Hg-199 EDM experiment, joined the He-6 beta decay experimental group. Elise Novitski – our first Robertson Postdoctoral Associate – joined CENPA and is working on Project 8 first tritium experiment. Megan Ivory began a postdoc working 50% on the Hg-199 EDM experiment. ADMX postdoc Rakshya Khatiwada accepted a research position at Fermilab.

Six graduate students received their Ph.D.'s in the last 12 months. They include Erik Lentz (University of Goettingen), Julieta Gruszko (MIT Pappalardo Fellowship), Matthias Smith (INFN Fellowship, Pisa), Brent Graner (CENPA Postdoc), Matt Turner (Microsoft Quantum Computing), and Eric Martin (seeking a Postdoc).

David Hyde retired in August, 2017. Still learning how to properly retire, David Hyde was rehired as the student shop manager on an hourly basis. Brittney Dodson has been hired to a new Research Engineer 1 position at CENPA. She begins in June, 2018.

Highlights

- In July 2017, KATRIN operated using a ^{83m}Kr source, obtaining precision ^{83m}Kr spectra and demonstrating readiness to begin tritium operation. On May 19th, the first tests with tritium gas were made, and a Curie plot was obtained demonstrating many important functioning aspects of the experiment.

- The TRIMS experiment at CENPA began running with tritium. Preliminary time-of-flight vs. ion energy distributions give qualitative support of modern molecular theory, and are in tension with much older experimental findings.

- The MAJORANA DEMONSTRATOR Collaboration published its first results on neutrinoless double-beta decay, obtaining a half-life limit of $T_{1/2} > 1.9 \times 10^{25}$y. They obtained a record energy resolution of 2.5 keV full width half maximum (FWHM) at $Q_{\beta\beta}$. A three-fold larger data set will be presented at the Neutrino 2018 conference.

- The LEGEND Collaboration was funded for basic R&D efforts. Initial plans are to deploy MAJORANA and GERDA Ge crystals together with additional contributions from the US and international partners to create the intermediate LEGEND-200 installation.

- The Project 8 tritium system has been built and is ready to inject tritium into a new CRES measurement cell. Plans continue in parallel for the Phase 3 large-bore scale up and the Phase IV production of atomic tritium.

- The COHERENT Collaboration published the first observation of coherent neutrino elastic scattering in Science, in August 2017. Additional NaI detectors are being prepared to extend the measurement program to light-nucleus targets.

- A new collaboration has been formed, and work has begun, to search for tensor currents measuring the beta spectrum of ^{6}He using the CRES technique pioneered by Project 8.

- The Muon $g - 2$ project completed a lengthy commissioning period and has begun its first physics data-taking campaign. The initial goal to slightly improve on the statistics of the BNL E821 experiment has been reached. The many UW-built systems are all working at or beyond design level.

- The analysis of the data from our MuSun experiment is reaching a very mature status and we hope to provide first results in 2018.

- The Hg-199 EDM experimental results have been re-analyzed to look for a time-varying EDM due to axion-like particles. New de-Gauss coils and new correction coils are prepared to reduce field gradients, which are the leading systematics in the most recent EDM campaign.

- The ADMX Gen2 experiment — located at CENPA — completed its first campaign reaching the DHFZ sensitivity goal, the only experiment in the world to have such an achievement. First results were published in PRL. Data taking is continuing.

- The CENPA LIGO Group is currently building four tiltmeters at LIGO Livingston Laboratory to improve seismic isolation and hence the duty cycle for gravitational-wave observation, after successful demonstration at LIGO Hanford Observatory last year.

As always, we encourage outside applications for the use of our facilities. As a convenient reference for potential users, the table on the following page lists the capabilities of our accelerators. For further information, please contact Gary Holman, Associate Director (holman@uw.edu) or Eric Smith, Research Engineer (esmith66@u.washington.edu) CENPA, Box 354290, University of Washington, Seattle, WA 98195; (206) 543 4080. Further information is also available on our web page: http://www.npl.washington.edu.

We close this introduction with a reminder that the articles in this report describe work in progress and are not to be regarded as publications nor to be quoted without permission of the authors. In each article the names of the investigators are listed alphabetically, with the primary author underlined in the case of multiple authors, to whom inquiries should be addressed.

David Hertzog, Director

Gary Holman, Associate Director and Editor
Charles Hagedorn and Daniel Salvat, Technical Editors
Ida Boeckstiegel, Administrative Editor

TANDEM VAN DE GRAAFF ACCELERATOR

Our tandem accelerator facility is centered around a High Voltage Engineering Corporation Model FN purchased in 1966 with NSF funds, with operation funded primarily by the U.S. Department of Energy. See W. G. Weitkamp and F. H. Schmidt, "The University of Washington Three Stage Van de Graaff Accelerator," *Nucl. Instrum. Methods* **122**, 65 (1974). The tandem was adapted in 1996 to an (optional) terminal ion source and a non-inclined tube #3, which enables the accelerator to produce high intensity beams of hydrogen and helium isotopes at energies from 100 keV to 7.5 MeV.

Some Available Energy Analyzed Beams

Ion	Max. Current (particle μA)	Max. Energy (MeV)	Ion Source
^{1}H or ^{2}H	50	18	DEIS or 860
^{3}He or ^{4}He	2	27	Double Charge-Exchange Source
^{3}He or ^{4}He	30	7.5	Tandem Terminal Source
^{6}Li or ^{7}Li	1	36	860
^{11}B	5	54	860
^{12}C or ^{13}C	10	63	860
*^{14}N	1	63	DEIS or 860
^{16}O or ^{18}O	10	72	DEIS or 860
F	10	72	DEIS or 860
* Ca	0.5	99	860
Ni	0.2	99	860
I	0.001	108	860

*Negative ion is the hydride, dihydride, or trihydride.

Several additional ion species are available including the following: Mg, Al, Si, P, S, Cl, Fe, Cu, Ge, Se, Br and Ag. Less common isotopes are generated from enriched material. We recently have been producing the positive ion beams of the noble gases He, Ne, Ar, and Kr at ion source energies from 10 keV to 100 keV for implantation, in particular the rare isotopes ^{21}Ne and ^{36}Ar. We have also produced a separated beam of 15-MeV ^{8}B at 6 particles/second.

Contents

INTRODUCTION   i

1 Neutrino Research   1
   KATRIN 1
   1.1 KATRIN status 1
   1.2 KATRIN detector operations 3
   1.3 KATRIN analysis software tools 5
   1.4 Study of the background from the KATRIN inter-spectrometer Penning trap   6
   1.5 KATRIN main spectrometer background from secondary electron emission   8
   TRIMS 9
   1.6 TRIMS overview and hardware progress 9
   1.7 TRIMS data and analysis framework 10
   MAJORANA 13
   1.8 Overview of MAJORANA DEMONSTRATOR and LEGEND 13
   1.9 Results of the MAJORANA DEMONSTRATOR 14
   1.10 Construction of a comprehensive model of radioactive backgrounds for the MAJORANA DEMONSTRATOR 17
   1.11 Multi-site background rejection in the MAJORANA DEMONSTRATOR 20
   1.12 Characterizing surface alpha events for MAJORANA and LEGEND 23
   1.13 Light readout with silicon photomultipliers for LEGEND 26
   1.14 Simulations and analysis for LEGEND 27
   1.15 Forward-biased preamplifiers for LEGEND 27
   Project 8 28
   1.16 Overview of Project 8 28
   1.17 Interpreting waveguide-effects on CRES signals 30
   1.18 A large-volume cavity CRES concept 32
   1.19 R&D towards atomic tritium production for Project 8 34
   COHERENT 37
   1.20 The COHERENT experiment: overview and first observation 37
   1.21 Crystal characterization and low-cost PMT base development for a ton-scale NaI array for COHERENT 39
   1.22 The COHERENT experiment: NaI simulations 40
   Selena 42
   1.23 Selena R&D 42

2 Non-accelerator-based tests of fundamental symmetries   45
   Torsion-balance experiments 45
   2.1 Fourier-Bessel gravitational inverse-square-law test 45

- 2.2 Progress on ground-rotation sensors for LIGO 46
- 2.3 An interferometric torsion-balance for atmospheric newtonian noise measurements 48
- 2.4 Preliminary limits on B-L coupled ultralight dark matter 49
- **Other tests of fundamental symmetries** 51
- 2.5 The mercury electric-dipole-moment experiment 51

3 Accelerator-based physics 53

- 3.1 Angular distribution of $2_1^+ \to 3_1^+$ photons in $^{21}\text{Ne}(p,\gamma)^{22}\text{Na}$ towards solving a puzzle 53
- 3.2 Overview of the ^{6}He experiments 54
- 3.3 Hardware improvements for the ^{6}He little-a experiment 56
- 3.4 Progress in the analysis of the ^{6}He little-a experiment 61
- 3.5 Progress towards precision measurement of the ^{6}He β-decay spectrum via cyclotron radiation emission spectroscopy 65

4 Precision muon physics 70

MuSun 70
- 4.1 Muon capture and the MuSun experiment 70
- 4.2 New data-driven method to quantify fusion-induced distortions in the MuSun experiment 73
- 4.3 MuSun 2015 data analysis 75
- 4.4 Measurements with the MuSun electron tracker 77
- 4.5 Monte Carlo framework and studies 79

Muon $g-2$ 81
- 4.6 Overview of the Muon $g-2$ experiment 81
- 4.7 Status of Muon $g-2$ experiment 85
- 4.8 Muon campus M4/M5 beamline optimization, transmission and MULTs ... 88
- 4.9 T0 detector, beam pulse shapes, and timing 91
- 4.10 Final installation of the calorimeters and gain sag mitigation 92
- 4.11 Energy-binned muon precession analysis 95
- 4.12 Performance of the calorimeter: timing and energy 99
- 4.13 IBMS update 102
- 4.14 Monitoring the $g-2$ storage ring magnetic field with fixed probes 105
- 4.15 Active shimming of the $g-2$ storage ring magnetic field using surface coils . 107

5 Dark matter searches 110

ADMX 110
- 5.1 Overview of ADMX 110
- 5.2 Higher-frequency axion searches with Orpheus 113

	DAMIC	116
	5.3 DAMIC: dark matter in CCDs	116
	5.4 CCD packaging and testing laboratory	118

6 Education — 120

6.1 Use of CENPA facilities in education and coursework at UW 120
6.2 Student training . . . 120
6.3 Accelerator-based lab class in nuclear physics . . . 123

7 Facilities — 126

7.1 Laboratory safety . . . 126
7.2 Van de Graaff accelerator and ion-source operations and development 130
7.3 Laboratory computer systems . . . 133
7.4 Electronic shop . . . 134
7.5 Instrument Shop . . . 137
7.6 Building maintenance, repairs, and upgrades . . . 142

8 CENPA Personnel — 143

8.1 Faculty . . . 143
8.2 CENPA external advisory committee . . . 143
8.3 Postdoctoral research associates . . . 144
8.4 Predoctoral research associates . . . 144
8.5 Undergraduates . . . 145
8.6 Visitors and volunteers . . . 146
8.7 Professional staff . . . 147
8.8 Technical staff . . . 147
8.9 Administrative staff . . . 147
8.10 Part-time staff and student helpers . . . 148

9 Publications — 149

9.1 Published papers . . . 149
9.2 Invited talks at conferences . . . 153
9.3 Abstracts and contributed talks . . . 155
9.4 Reports, white papers and proceedings . . . 155
9.5 Ph.D. degrees granted . . . 156

1 Neutrino Research

KATRIN

1.1 KATRIN status

J. F. Amsbaugh, A. Beglarian*, T. Bergmann*, T. H. Burritt, P. J. Doe, S. Enomoto, J. A. Formaggio[†], F. M. Fränkle*, M. Howe[‡], L. Kippenbrock, A. Kopmann*, Y.-T. Lin, N. S. Oblath[§], D. S. Parno[¶], D. A. Peterson, A. W. P. Poon[‖], R. G. H. Robertson, D. Tcherniakhovski*, L. Thorne[¶] A. Vizcaya[¶] T. D. Van Wechel, J. F. Wilkerson[‡], and S. Wüstling*

The KATRIN experiment will probe the neutrino mass to an expected sensitivity of 0.2 eV (90% C.L.) using the well-established technique of precision tritium beta-decay electron spectroscopy. KATRIN has taken the technology of a windowless-gaseous tritium source, coupled to a retarding-potential spectrometer, to its practical limit. Seventeen years in the making, the highlight for 2017 was a demonstration that the sub-components of KATRIN function in concert and are ready to begin tritium operation.

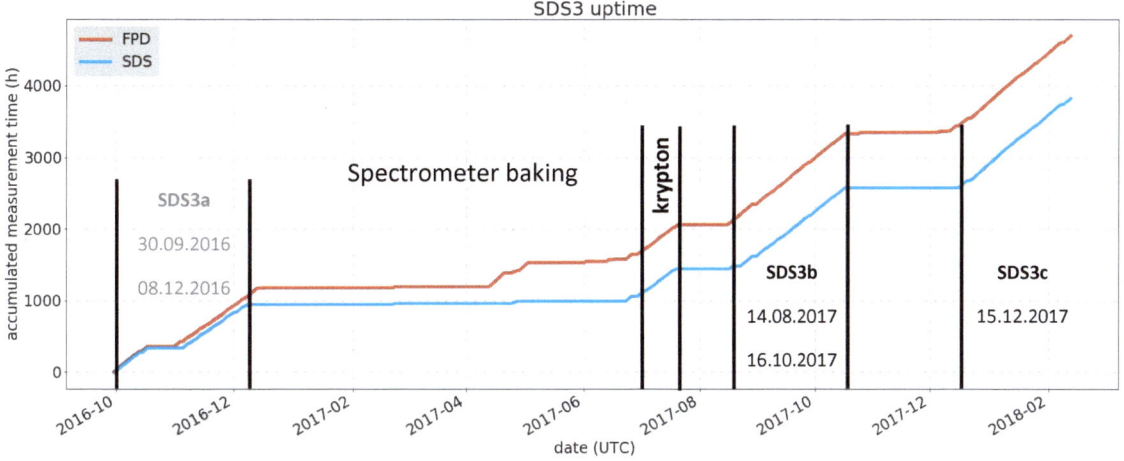

Figure 1.1-1. Significant activities of 2017, showing the spectrometer baking, the krypton run and the two SDS campaigns. The detector system accrued approximately 3,600 hours of operation, or about 40% livetime, without any major problems.

The activities of the past year, seen in Fig. 1.1-1, consisted of the completion of the third in

---
*Karlsruhe Institute of Technology, Karlsruhe, Germany.
[†]Massachusetts Institute of Technology, Cambridge, MA.
[‡]University of North Carolina, Chapel Hill, NC.
[§]Pacific Northwest National Laboratory, Richland, WA.
[¶]Carnegie Mellon University, Pittsburg, PA.
[‖]Lawrence Berkeley National Laboratory, Berkeley, CA.

a series of measurements using the Spectrometer-Detector-System (SDS3) and the operation of the entire KATRIN instrument with ^{83m}Kr. It was shown during initial commissioning tests that the primary background in the main spectrometer was due to ^{210}Pb on the inner wall of the main spectrometer, resulting in a background rate five times the design target of 10 mHz. The ^{210}Pb originated from the decay chain of airborne radon. In addition, a Penning trap, known to exist between the two spectrometers was shown to be a significant problem. The initial commissioning tests were conducted at a spectrometer pressure of 10^{-9} mbar. Prior to Kr operation and to achieve the design pressure of 10^{-11} mbar, the spectrometers were baked. Under these improved vacuum conditions, subsequent commissioning showed that the Penning trap no longer discharged. Using a high power laser, the "light hammer", an attempt was made to ablate the ^{210}Pb from the spectrometer walls. Unfortunately this did not reduce the background, which remains at 50 mHz. A subsequent test was conducted during the 14-day "quiet" New Year period. With the spectrometers under nominal tritium operation conditions it was shown that the background rates were stable, there were no Penning discharges and scanning the analyzing voltage of the spectrometer did not affect the background rate. The background was found to be energy independent at the end-point. A slope in the background would introduce potential systematic effects in the neutrino mass analysis.

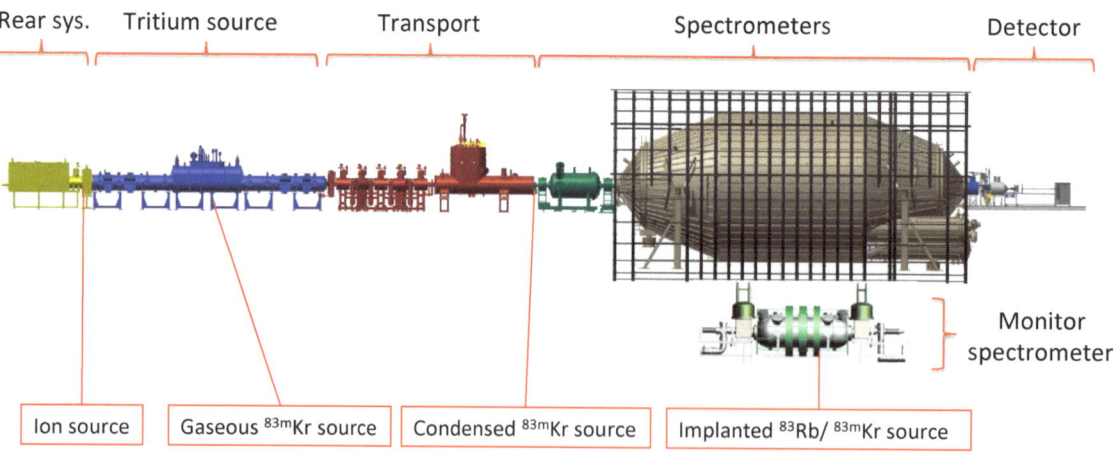

Figure 1.1-2. Location of the Kr sources in the KATRIN apparatus.

Operating the entire KATRIN apparatus using ^{83m}Kr sources at various locations in the beam-line exercised the hardware, data acquisition plan, and the analysis software; a major milestone in the road to tritium operation. Fig. 1.1-2 shows the locations of the three ^{83m}Kr sources deployed. ^{83m}Kr is an ideal source for testing low-energy, low-background apparatus. It has well-defined internal conversion lines at 17 and 32 keV. Furthermore, its half-life of 1.8 hr means that any accidental contamination is not long-lasting, while the 86 d half-life of the mother, ^{83m}Rb, makes it a practical source for long term investigations. ^{83m}Kr was injected at a steady rate into the source tube. The conversion lines allowed study of an isotropic distribution of electrons in the source tube, mimicking a tritium source and its gas dynamics. In order to avoid condensing the Kr gas, the source was operated at a temperature of 100 K. Tritium operation requires operation of the source at 30 K, excluding the possibility of simultaneous operation with Kr. To overcome this difficulty, a condensed Kr source can be deployed just upstream of the pre-spectrometer. The third Kr source consists of implanted ^{83m}Rb ions in a highly oriented pyrolytic graphite foil, mounted in the auxiliary "monitor spectrometer". Using the monitor spectrometer it was observed that the Kr conversion lines are stable over a typical three-month KATRIN data taking cycle to a part in 10^6. Since the main spectrometer and the monitor spectrometer share the same retarding potential power supply, monitoring the line position is equivalent to measuring the stability of the retarding potential of the spectrometer. It is critically important that tritium ions produced in the source, or by scattering in the beam-line, do not enter and contaminate the main spectrometer. Such contamination would create a new background. Using a source of deuterium ions it was demonstrated that electrodes installed in the beam-line, in conjunction with the cryogenic pumping system and the appropriate potential on the pre-spectrometer, prevent all ions from entering the main spectrometer. Using these sources and associated detectors we have been able to demonstrate that the stability and performance of the KATRIN subcomponents meets and in some cases exceeds specification. KATRIN is therefore ready for first tritium operation scheduled for June 2018. The gaseous source will initially contain approximately 1% tritium with the balance made up of deuterium. Ramping up to the full source strength of $\sim 10^{11}$Bq will commence in October 2018.

1.2 KATRIN detector operations

P. J. Doe and L. Kippenbrock

The detector system performed reliably over 2017, accruing approximately 3,600 hours of operation during the Krypton and Spectrometer-Detector-System (SDS) campaigns. During data-taking, the FPD (Focal Plane Detector) is regularly calibrated with an ^{241}Am source, which produces gammas at fixed energies[1]. These calibration data can be used to study the energy resolution of the detector. Fig. 1.2-1 shows the energy resolution as a function of time. While the resolution appears generally stable during a particular measurement campaign, two energy resolution degradations of greater than ten percent are apparent. No definitive explanation for these degradations in the energy resolution has been found, although there are indications that the issue lies with the wafer and not the electronics read-out chain. The more recent degradation in the energy resolution seems to be an effect of wafer damage from specific high rate-measurements, although the mechanism for the degradation is not understood. Additional studies of the detector wafers are needed.

---

[1] J.F. Amsbaugh *et al.* Nucl. Instrum. Methods A **778**, 40-60 (2015).

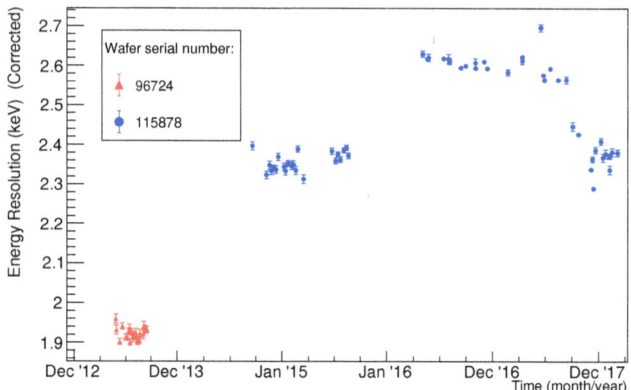

Figure 1.2-1. The energy resolution of the FPD as a function of time, as determined from fitting the 60-keV line of the ^{241}Am-source data. The energy spectra for all pixels were summed together before fitting. Two different wafers have been installed in the system for measurement campaigns, indicated by the different markers. To correct for the known temperature-dependence of the energy resolution, an approximate and preliminary linear correction has been made to each data point based on the system temperature, where the normalization temperature is 0° C.

This unexplained detector variation highlighted the need for an independent apparatus with which to test wafers since the FPD system will be continuously busy taking KATRIN tritium data. Fortunately, a duplicate set of data acquisition electronics, the so-called "Iron Bird", equipped with a dummy detector wafer, was available for use. A vacuum chamber was constructed to house an actual detector wafer and front-end electronics, cooled by a Peltier cooler. This apparatus can be seen in Fig. 1.2-2. Once commissioned, the Iron Bird will not only allow detector wafers to be qualified but also serve as a test bench for further development of the DAQ system to accommodate the high data rates associated with searches for sterile neutrinos.

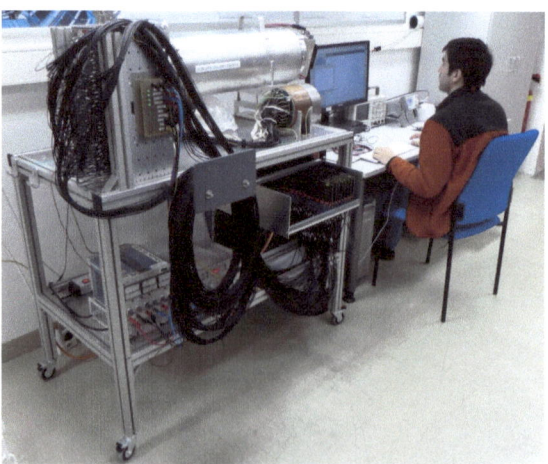

Figure 1.2-2. The Iron Bird DAQ and detector wafer test apparatus.

1.3 KATRIN analysis software tools

S. Enomoto

The UW CENPA group is responsible for the KATRIN data-analysis tools and has provided virtually everything in the data-analysis tool-chain (totaling more than 100,000 lines of code in 650 files). Fig. 1.3-1 shows an overall view of the KATRIN analysis tools. The software components with a red label are provided by the UW CENPA group.

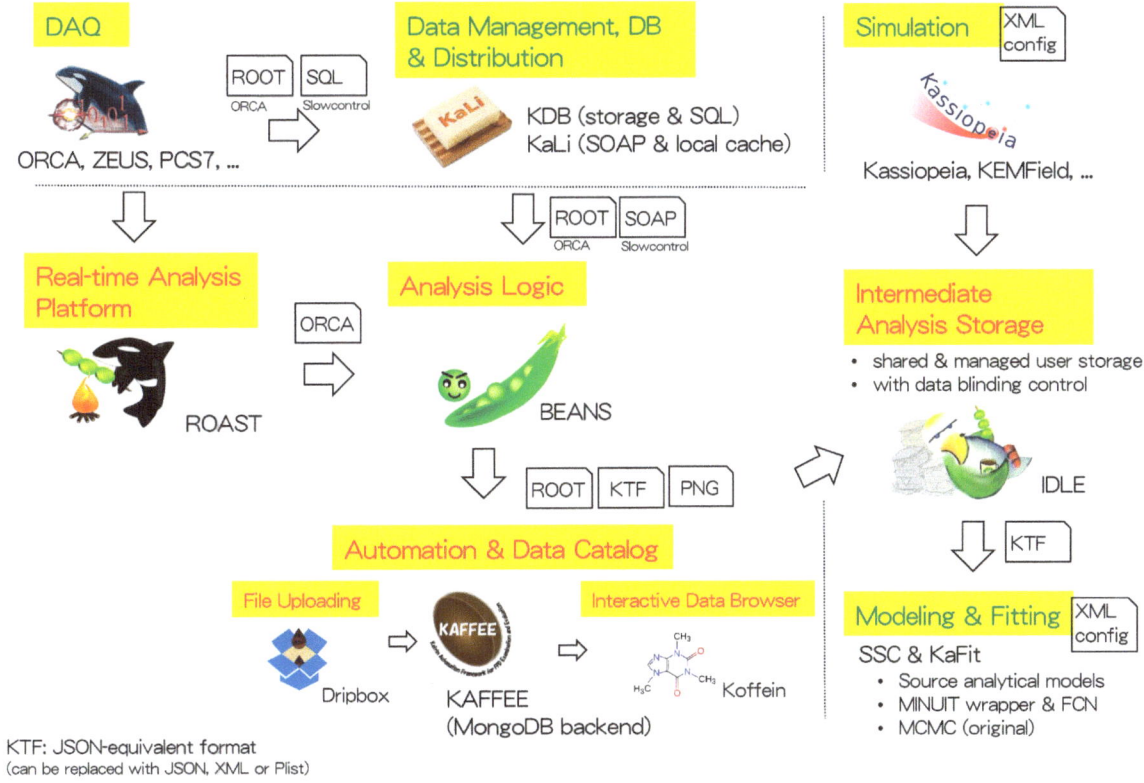

Figure 1.3-1. Overview of the KATRIN analysis tools. Components with a red label are provided by the UW CENPA group.

The core of the data-analysis tool-chain is BEANS (Building Elements for ANalysis Sequence), an analysis platform and mechanism with which users can construct analysis logic *without programming*. It was originally developed for KATRIN commissioning analysis where short time-to-result and adaptability to unpredicted analyses were important. It was found later that these features were also beneficial for offline analysis, and BEANS was adopted for KATRIN standard analysis software. BEANS was designed with use for real-time analysis (e.g., growing histograms for data broadcasting from ORCA) in mind; now we can use a single BEANS script for both offline analysis and real-time analysis, enabling a full analysis in real-time (with ROAST – Realtime Orca Analysis on STreaming-data). We also provided an automation framework (KAFFEE – Katrin Automation Framework for FPD Examination and Evaluation) for near-time monitoring and run cataloging, with browser-based interactive analysis (Koffein) for data on the catalog.

In 2017, we operated the entire KATRIN apparatus with a gaseous ^{83m}Kr source. It was the first measurement involving all the major components of KATRIN, such as the gaseous source at 100 K and electron transport from the source section to the detector through the spectrometers. It was also the first opportunity for us to integrate all the software components, including data analysis, simulation (of electron tracking etc), and fitting of physics models (such as the ^{83m}Kr spectrum). The UW CENPA group provided an overall design for software integration, with a key component of integration, IDLE, an intermediate storage with additional facilities such as data quality monitoring and data blinding.

All the data from the 2017 krypton runs were transferred to UW CENPA automatically by KAFFEE to be processed with a set of BEANS scripts. The digested run files were then put on IDLE and distributed to all the KATRIN members. During data taking, Koffein was heavily utilized to interactively explore the BEANS-preanalyzed results on the KAFFEE run catalog at CENPA. The same data-analysis scheme will be used for the tritium runs that begin in May 2018.

1.4 Study of the background from the KATRIN inter-spectrometer Penning trap

M. Fedkevych[*], F. Fränkle[†], L. Kippenbrock, D. S. Parno[‡], and P. C.-O. Ranitzsch[*]

In its designed configuration, the KATRIN main spectrometer is placed on negative high voltage in order to act as a high-resolution MAC-E filter (Magnetic Adiabatic Collimation with an Electrostatic Filter) for electrons produced from tritium β-decay. In order to reduce background events inside the main spectrometer that result from scattering on residual gas, the pre-spectrometer can also be placed on negative high voltage (at a value slightly more positive than that of the main spectrometer), in order to act as a pre-filter for the electrons and thus reduce the flux of low-energy electrons into the main spectrometer. However, this setup creates a Penning trap for electrons, formed by the two negatively-charged spectrometers and the confining superconducting magnet between the spectrometers. Trapped electrons will scatter on residual gas, producing positive ions that can be accelerated into the main spectrometer. Inside the main spectrometer, these ions can ionize residual gas and produce ionization electrons that can reach the KATRIN focal-plane detector and appear as background. The creation of background events from the build-up of electrons in the trap is referred to as a "Penning discharge".

It is imperative to have a low background rate in order to achieve KATRIN's design sensitivity on the effective neutrino mass. Thus, the contribution of the Penning trap to the main spectrometer background must be experimentally measured and mitigated, if the pre-spectrometer is to be used as a pre-filter. Previous measurements with the Penning trap in

---

[*]University of Münster, Münster, Germany.
[†]Karlsruhe Institute of Technology, Karlsruhe, Germany.
[‡]Carnegie Mellon University, Pittsburgh, PA.

2016[1] saw large and unavoidable Penning discharges, even at low pre-spectrometer voltages, although these measurements were performed at relatively high residual gas pressures.

In September 2017, two weeks of dedicated Penning trap measurements were performed with close-to-nominal pressure ($\sim 10^{-11}$ mbar) inside the main spectrometer. The conclusion from these measurements was that the design voltage for the pre-spectrometer could be achieved without the continuous Penning discharges that plagued earlier measurements. To study the effect of pressure in more detail, background measurements with the Penning trap were also performed at an elevated pressure, by injecting argon gas into the main spectrometer. A clear pressure dependence to the Penning trap-induced background rate was observed, which can be seen in Fig. 1.4-1. This result indicates that KATRIN can mitigate the effect of the Penning trap by maintaining ultra-high vacuum conditions inside the main spectrometer.

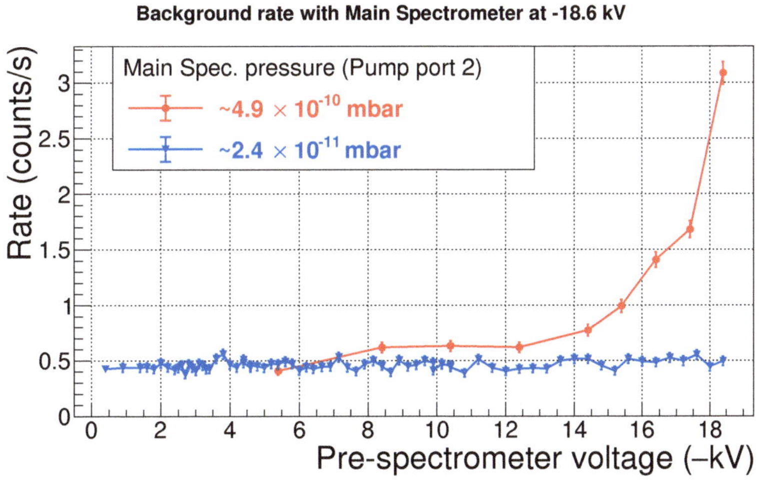

Figure 1.4-1. The background rate observed inside the main spectrometer as function of the pre-spectrometer voltage. Data were collected at high and low pressures, indicated by the red and blue points, respectively. The flat distribution for the low-pressure data indicates that the Penning trap does not significantly contribute to the background rate at low pressures, unlike at higher pressures where there is a large rise in the rate as the pre-spectrometer voltage increases (i.e. as the trap deepens).

KATRIN performed a long-term (>2 weeks) background measurement starting in December 2017 with both spectrometers placed on high voltage. No Penning discharges were observed. Furthermore, comparing the background rates from the main spectrometer with and without the pre-spectrometer on high voltage, no significant rate contribution from the Penning trap was observed. Further background studies with the Penning trap are planned to see the effect of high fluxes of electrons into the pre-spectrometer, to simulate conditions during tritium operation. The high flux of electrons could increase the number of particles in the Penning trap and may thereby result in an elevated background rate.

---

[1]CENPA Annual Report, University of Washington (2017) p. 5.

1.5 KATRIN main spectrometer background from secondary electron emission

F. Fränkle*, L. Kippenbrock, D. S. Parno[†], P. C.-O. Ranitzsch[‡], and J. Wolf*

Cosmic-ray muons and environmental gamma radiation will produce secondary electrons when passing through the steel walls of the main spectrometer. Due to the spectrometer's large size, background events from these secondary electron sources can be significant if left unshielded. KATRIN shields the flux tube from secondary electrons by use of the magnetic flux tube's inherent self-shielding properties (i.e. Lorentz force), as well as a two-layer wire electrode system installed near the vessel walls that reflects charged particles. The adequacy of this shielding system must be experimentally verified.

The muon-induced background has been studied with several muon detectors – scintillator panels installed close to the main spectrometer. In particular, a correlation analysis can be performed by comparing the FPD (Focal-Plane Detector) electron rate and the muon detector rate. For an electromagnetic configuration similar to the nominal KATRIN operating mode, no correlation between the FPD and muon rates was observed. As previously reported[1], a paper describing the analysis of the muon-induced background is in preparation. In the last year, several improvements have been made to the analysis. A new statistical "toy" model was developed to estimate the sensitivity of the muon-FPD correlation experiments. Furthermore, a more-sophisticated correlation-fitting scheme was developed, which permits a more-restrictive upper limit on the muon-induced background rate. In addition, we were able to use the correlation data to measure the secondary electron production rate for muons passing through the main spectrometer surface.

The effect of gamma radiation has been investigated using a 53 MBq ^{60}Co source placed near the main spectrometer. When measured with magnetic and electric shielding enabled, no effect on the background rate was observed from the ^{60}Co source. To compute an upper limit on the background contribution from environmental gammas, a comparison is made to a GEANT4[2] simulation of the gamma flux inside the main spectrometer. The simulation can also be used, in combination with measurements made with other electromagnetic settings, to probe the fraction of secondary electrons emitted from the main spectrometer that are gamma-induced. This background analysis is also being prepared for publication.

---

*Karlsruhe Institute of Technology, Karlsruhe, Germany.
[†]Carnegie Mellon University, Pittsburgh, PA.
[‡]University of Münster, Münster, Germany.

[1]CENPA Annual Report, University of Washington (2017) p. 6.
[2]S. Agostinelli *et al.*, Nucl. Instrum. Methods A **506**, 250 (2003).

TRIMS

1.6 TRIMS overview and hardware progress

W.-J. Baek[‡], T. H. Burritt, C. Claessens[†], S. Enomoto, G. Holman, M. Kallander, Y.-T. Lin, E. Machado, R. Ostertag[‡], D. S. Parno[§], J. R. Pedersen, D. A. Peterson, R. G. H. Robertson, E. B. Smith, T. D. Van Wechel, A. P. Vizcaya Hernandez[§], and D. I. Will

The goal of the Tritium Recoil-Ion Mass Spectrometer (TRIMS) experiment is to measure the branching ratios for decays of molecular tritium (T_2) into bound and unbound states. The experiment uses coincident signals with an electron arriving at one end of an acceleration chamber in a uniform magnetic field, and ions arriving at the other end. From the timing difference as well as the detected energy, different ion species can be distinguished so as to obtain the branching ratio in question.

In the original implementation of the TRIMS apparatus, the decay chamber was a 20 cm long Pyrex tube. This configuration was more than sufficient to hold 60 kV, but in practice, electrons freed from the surface of the glass are accelerated part way down the chamber and strike the glass again, releasing a cascade of electrons. As the electron cascades reached the high voltage end, the glass was left more positively charged, attracting still more secondary electrons. Eventually the potential difference is enough to cause a surface discharge. Both the cascades and the discharges made light and produced significant background in the Passivated Implanted Planar Silicon (PIPS) detectors used in the experiment.

The Pyrex chamber has been replaced with an acceleration column from National Electrostatics Corporation, as shown in Fig. 1.6-1. From the inner surface of the column's ceramic spacers, there is no path for electrons to travel along the ceramics without soon striking an electrode, and there is no line of sight for the detectors to see any light that could emanate from secondary electrons emitted by the column itself.

[‡]Karlsruhe Institute of Technology, Karlsruhe, Germany.
[†]Johannes Gutenberg University, Mainz, Germany.
[§]Carnegie Mellon University, Pittsburgh, PA.

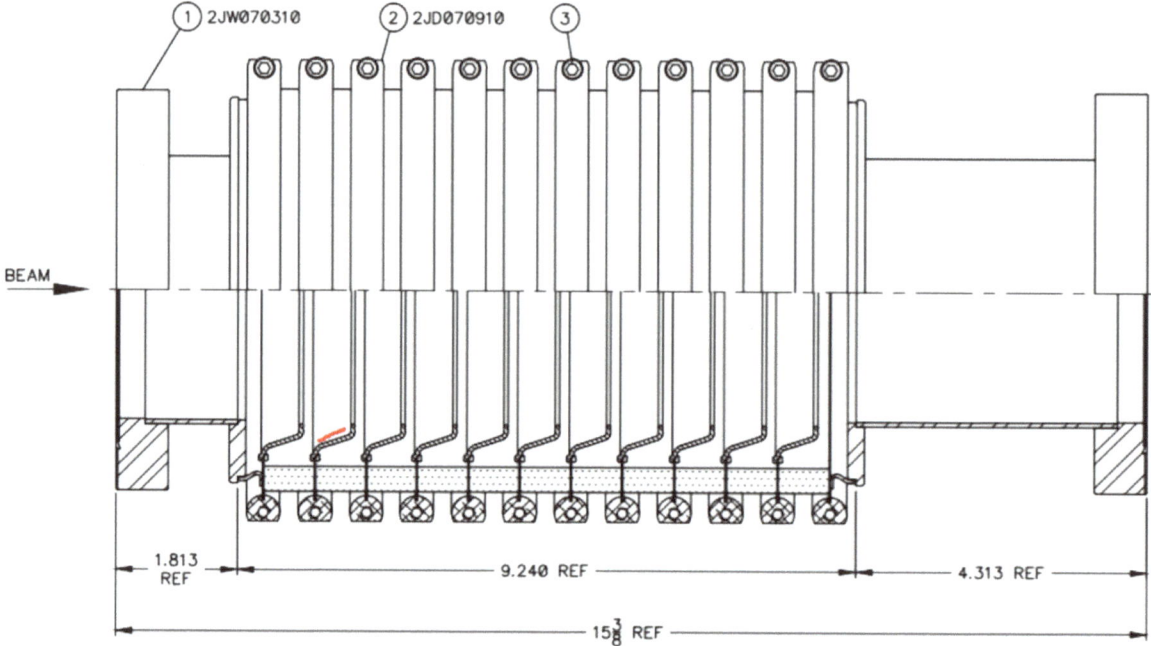

Figure 1.6-1. A cut-away schematic diagram of the new TRIMS acceleration column for National Electrostatics Corp. In this picture, the ion detector will be at the right side, and the beta will be at the left. The S-shaped cross section of the ring-electrodes suppresses secondary electrons from the ceramic spacers.

In addition to the new chamber, a translation-stage for horizontal as well as vertical movement of the ion detector has been added. This allows us to quantify the error associated with misalignment relative to the beta detector, as well as allowing finer detector steps that can be useful for the deconvolution analysis.

Moreover, Joule-Thomson coolers have been designed to cool the detectors in order to reduce the detector noise caused by leakage current. With these new devices, an energy resolution of 2 keV full width half maximum (FWHM) has been achieved on both detectors.

With the chamber replaced, the high-voltage breakdowns have ceased, and TRIMS has been taking T_2 and HT (tritium hydride) data with great success. A publication on the HT data is in preparation.

1.7 TRIMS data and analysis framework

M. Kallander, Y.-T. Lin, E. M. Machado, D. S. Parno[*], R. G. H. Robertson, and A. P. Vizcaya Hernandez[*]

TRIMS began taking data soon after the installation of the new chamber. The preliminary data from T_2 mixed with HT are shown in Fig. 1.7-1, where the x-axis indicates the relative

---
[*]Carnegie Mellon University, Pittsburgh, PA.

time of flight and y-axis the ion detected energy. The four main quadratic bands correspond to the ions of mass 1, 3, 4, 6 (amu) as the expected beta decay products. Secondary features such as doubly-charged ions have also been identified.

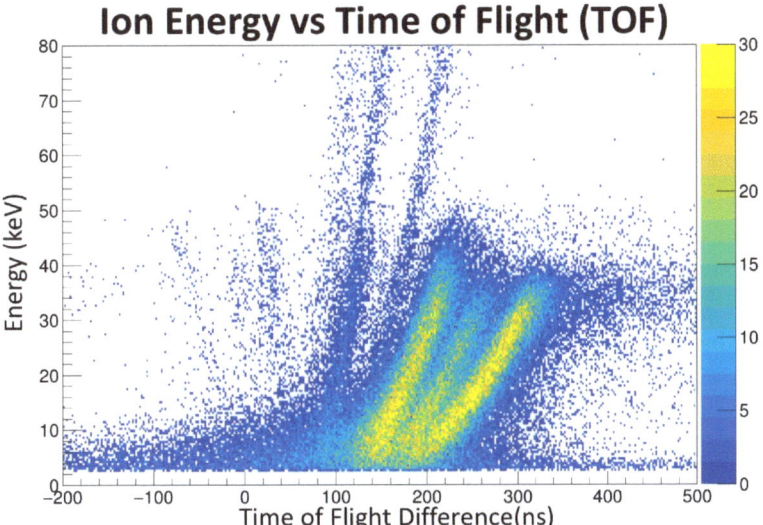

Figure 1.7-1. Preliminary histogram from the TRIMS experiment showing decays from HT and T_2. The horizontal axis shows the difference in detection times for each beta-particle and the corresponding ion. The vertical axis shows the ion kinetic energies observed by the ion detector. The color axis indicates the number of counts in each histogram bin. From left to right, the yellow-colored bands are the mass 3, mass 4, and mass 6 ions, respectively.

In addition, TRIMS has improved in several areas of analysis, including the detector timing resolution. Previously, the trigger time of an event waveform was obtained by the zero-crossing of a double-trapezoidal filter. The new method uses a digital constant fraction discriminator (dCFD) – the trigger time is defined as the point when the signal pulse rises to a fixed fraction of the overall pulse amplitude. Moreover, a Woods-Saxon function, also known as the logistic function, is fitted to the waveform before dCFD. The fitting procedure reduces the impact of electronic noise, which is significant in the keV detection range. As a result, the detector timing resolution of TRIMS has been improved by roughly an order of magnitude.

We have updated our simulation: The dead-layer effect of the silicon detectors is considered using the SRIM[1] (The Stopping and Range of Ions in Matter) software package, which generates output that can be used by our GEANT4 simulation. In addition, a new software tool has been developed using SRIM to evaluate the energy loss, the energy broadening by straggling, and the backscattering rate over wide ranges of ion incident energy and dead layer width. This update allows us to identify mass-bands in TRIMS data by comparison with simulation.

The quantities of interest to the TRIMS experiment are the branching ratios to the dif-

---

[1] J. F. Ziegler, M. D. Ziegler, and J. P. Biersack, Phys. SRIM, Ion Implantation Press (2010).

ferent ion species. The full transverse extent of the ions' cyclotron orbits as they traverse the region of uniform electric and magnetic fields is substantial, and must be taken into account to accurately measure the branching ratios. At the ion detection plane, the distribution of possible locations of ions, called the data-density function (DDF) is larger than the ion detector's active area. The branching ratios are computed by taking the integrals of these ion-specific DDFs, and normalizing them by the total number of decays detected. To obtain the DDFs, we measure "raw-count functions" (RCFs) by translating the ion detector radially outward from the axis in fixed intervals and counting the number of each ion species detected. A DDF of annular bins can be transformed to an RCF using a point-spread function (PSF). The PSF is calculated analytically by overlapping the circular ion detector with the DDF and counting the fraction of each annular bin area that is covered by the ion detector, as shown in Fig. 1.7-2. To compute the DDF from the RCF, a least-squares analysis is applied and will be well-behaved, provided there is not much noise on the RCF. Tests using a simulated DDF confirm that the error introduced is less than 1% when the total counts of the RCF exceeds one million, which is comparable to the size of TRIMS' current data set.

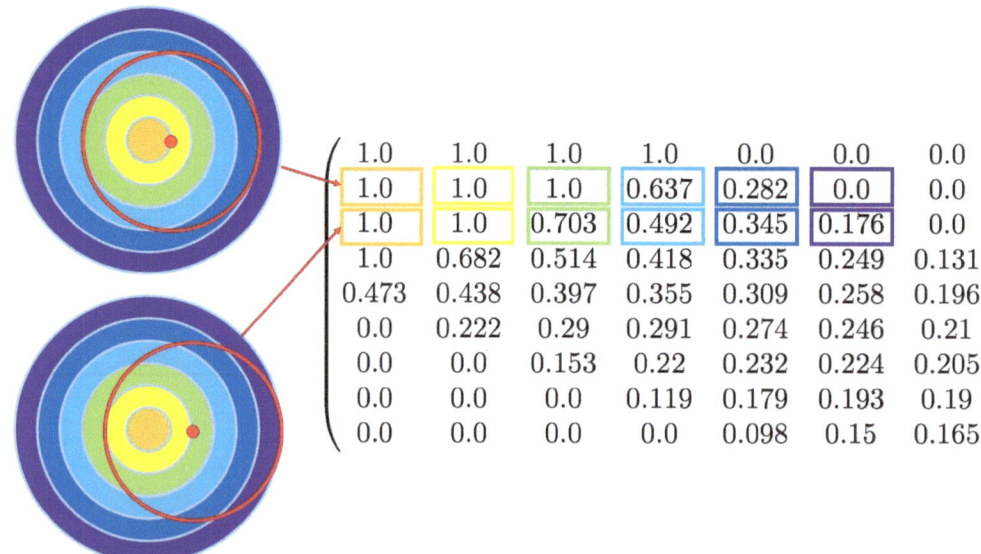

Figure 1.7-2. Our method for constructing the point-spread function (PSF) for the TRIMS analysis, depicted for two different radial positions of the detector. The red circle represents the area covered by the ion detector. Each colored annulus corresponds to an annular bin of the expected data-distribution function (DDF). Rows of the matrices correspond to increasing radial translation of the detector. Columns of the matrices correspond to the colored annular rings. The values within the matrices are the fraction of each annular bin that is covered by the detector.

MAJORANA

1.8 Overview of MAJORANA DEMONSTRATOR and LEGEND

S. I. Alvis, T. H. Burritt, M. Buuck, C. Cuesta*, J. A. Detwiler, Z. Fu[†], J. Gruszko[†], I. Guinn, D. A. Peterson, W. Pettus, N. W. Ruof, and T. D. Van Wechel

The MAJORANA DEMONSTRATOR is a neutrinoless double-beta ($0\nu\beta\beta$) decay search employing a 44.1 kg array of high-purity germanium (HPGe) detectors, 29.7 kg of which are enriched in ^{76}Ge. $0\nu\beta\beta$ is a presently-unobserved rare process that would violate lepton number conservation and indicate that the neutrino is a Majorana fermion[1]. The experiment is in operation in the Davis Campus at the 4850-foot level of the Sanford Underground Research Facility (SURF) in Lead, SD. The detectors are arranged in linear "strings" of 3 to 5 on low-background underground electroformed copper supports which are then placed into two cryogenic modules inside a passive shield, as shown in Fig. 1.8-1. The primary technical goal of the DEMONSTRATOR is the demonstration of a radioactive background level of 3 counts/ton/year within a 4 keV region of interest (ROI) surrounding the 2039 keV Q-value for ^{76}Ge $0\nu\beta\beta$ decay. Reaching such a low background level would justify deeper investment in a larger tonne-scale experiment with sufficient sensitivity to definitively search for $0\nu\beta\beta$ decay for inverted-hierarchical neutrino masses. Recent results from the DEMONSTRATOR will be summarized in (Sec. 1.9).

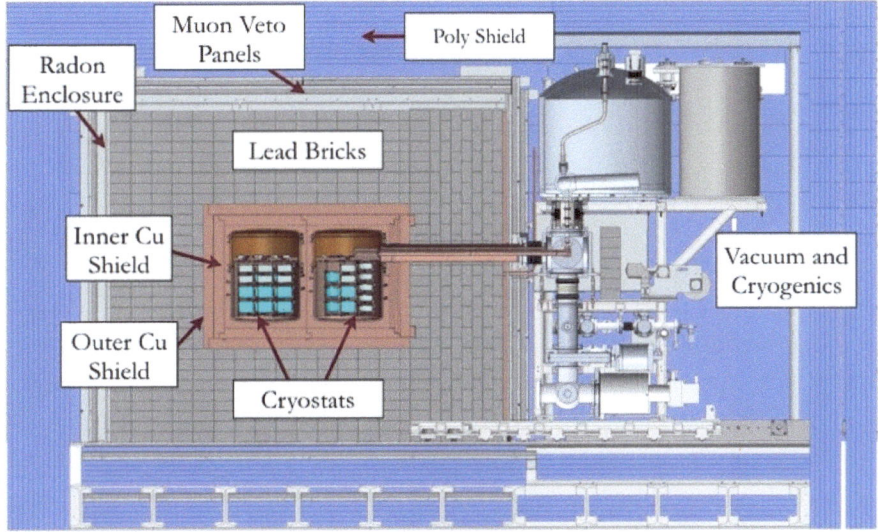

Figure 1.8-1. Schematic of the MAJORANA DEMONSTRATOR apparatus. Two independent modules, totaling 44 kg of p-type point contact (PPC) germanium detectors, are operated inside a layered passive shield. The shielding layers from the inside are: low-background underground electroformed copper, commercial copper, lead, and polyethylene.

---

*Presently at Centro de Investigaciones Energeticas, Medioambientales y Tecnologicas, CIEMAT 28040, Madrid, Spain.
†Presently at Massachusetts Institute of Technology, Cambridge, MA.

[1]N. Abgrall *et al.*, Adv. High Energy Phys. **2014**, 365432 (2014).

In order to perform a tonne-scale ^{76}Ge $0\nu\beta\beta$ search, the Large Enriched Germanium Experiment for Neutrinoless $\beta\beta$ Decay (LEGEND) collaboration has been formed. Building on the successes of GERDA[1] and the MAJORANA DEMONSTRATOR, a phased experimental program is envisioned to achieve this goal[2]. LEGEND-200 will deploy 200 kg of HPGe detectors, including enriched detectors from both GERDA and the MAJORANA DEMONSTRATOR, in the GERDA liquid-argon veto at the Laboratori Nazionali del Gran Sasso (LNGS) in Italy. LEGEND-200 will improve the half-life sensitivity by an order of magnitude to $> 10^{27}$ yr and decrease the background by a factor of 5 over the present generation of experiments to 0.6 counts/FWHM/t/yr. Operation will begin in 2021, utilizing existing infrastructure. LEGEND-1000 will target an ultimate sensitivity greater than 10^{28} yr, covering the inverted-hierarchy allowed-region, by deploying a tonne of HPGe detectors and achieving a background of 0.1 counts/FWHM/t/yr.

In support of the requirements of LEGEND, CENPA is conducting several targeted R&D projects. Studies of light collection in liquid argon, beginning with silicon photomultiplier (SiPM) characterization, are underway using equipment from the LArGe test stand[3] (Sec. 1.13). An investigation into alternative front-end electronics is reviving previous work with a forward-bias reset design[4] (Sec. 1.15).

1.9 Results of the MAJORANA DEMONSTRATOR

S. Alvis, T.H. Burritt, M. Buuck, J. Detwiler, Z. Fu*, J. Gruszko*, I. Guinn, D.A. Peterson, W. Pettus, R.G.H. Robertson, N. Ruof, and T.D. Van Wechel

The MAJORANA DEMONSTRATOR has been taking data with one or both modules since June 2015. Data taking is divided into datasets based on changes to the DAQ or hardware configuration. A summary of the contents of each dataset is contained in Fig. 1.9-1. Dataset 0 includes Module-1 data from June 27 to Oct. 7, 2015, before the inner copper shield was installed. Dataset 1 began after completion of the inner copper shield and lasted from Dec. 31, 2015 to May 24, 2016. Dataset 2 used the multisampling feature of the GRETINA digitizers, which downsamples the baseline and falling tail of each waveform while sampling rising edge at the full frequency in order to increase the sampling length of the waveform without compromising the pulse-shape analysis. Dataset 2 lasted from May 24 to July 14, 2016. Module 2 was installed in July, 2016, and datasets 3 and 4 were taken concurrently from Aug. 25 to Sept. 27, 2016. For these datasets, the module 1 and module 2 DAQ systems were run separately, with module 1 data in dataset 3 and module 2 in dataset 4. The DAQ systems were integrated in Oct., 2016, and dataset 5 was taken from Oct. 13, 2016 to May 11, 2017. Dataset 5 was split into three subdatasets. Dataset 5a had elevated noise due to poor

---

*Presently at Massachusetts Institute of Technology, Boston, MA.

[1]M. Agostini *et al.*, Phys. Rev. Lett. **120**, 132503 (2018).
[2]N. Abgrall *et al.*, AIP Conf. Proc. **1894**, 020027 (2017).
[3]CENPA Annual Report, University of Washington (2007) p. 27.
[4]CENPA Annual Report, University of Washington (2015) p. 25.

high voltage grounding. Dataset 5b began on Jan. 27 with optimized grounding. Dataset 5c began after 10 kg yr of total unblinded physics exposure had been taken, on Mar. 17, 2017. Dataset 6 began May 11, 2017 when multisampling was enabled on both modules, and is currently ongoing. As of April 15, 2018, approximately 30 kg yr of combined blinded and unblinded physics data have been taken.

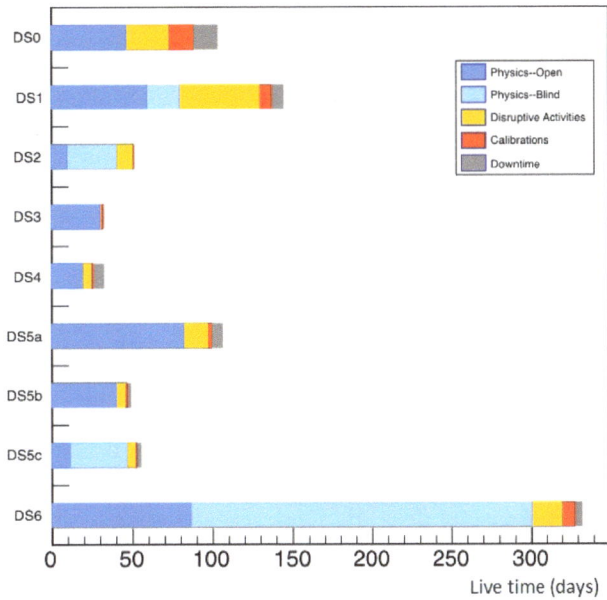

Figure 1.9-1. Summary of the duty cycle for the DEMONSTRATOR as of April 15, 2018.

Detector calibrations of each module are performed once per week by exposing the detectors to a ^{228}Th line source for 90 minutes. The source is inserted and removed by motor into guide tubes that wrap around each module. Energy calibration and measurement of detector resolution are performed using the γ-peaks, excluding the 511-keV electron annihilation peak and the 1592-keV and 2103-keV pair production peaks. Energy calibration is performed individually for each detector between each week by gain-matching the 2614-keV and 238-keV peaks. For each dataset, a simultaneous fit of peaks is performed using an analytic peak-shape model consisting of a Gaussian, a low energy exponentially modified Gaussian (exgaus) tail, a high energy exgaus tail and a step consisting of an erfc function. This fit is used to measure a correction to the energy calibrations that ensures that the calibrations between each detector matches accurately. The peak-shape parameters are used to determine the optimal ROI and the full width half maximum (FWHM) as a function of energy of the combined array. The FWHM was measured to be 2.5-keV at the 2039-keV $Q_{\beta\beta}$-value. The mean optimal ROI over all datasets was 4.32 keV. The calibration spectrum and FWHM function are shown in Fig. 1.9-2.

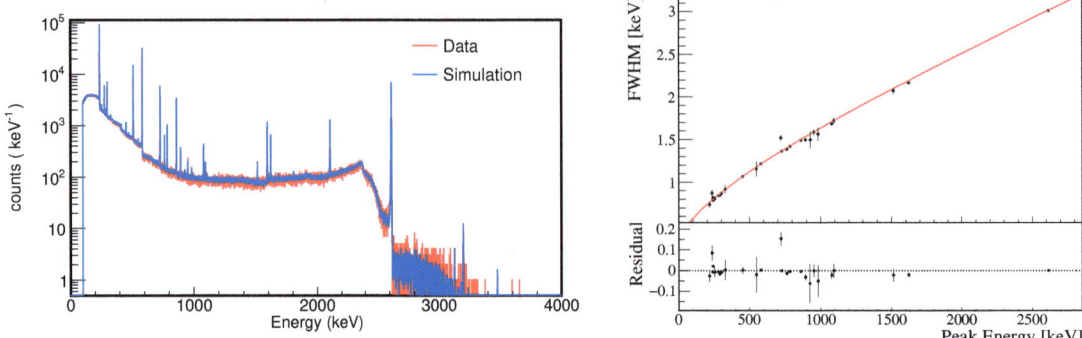

Figure 1.9-2. *Left:* Calibration spectrum. Calibrations are performed using between 8 and 35 peaks, depending on the number of calibration runs included. *Right:* the FWHM of the combined spectrum of all detectors measured as a function of energy.

Multiple data filters are applied to reduce backgrounds. First, several data cleaning routines cut non-physical waveforms. Waveforms from different detectors that occur within 4 μs of each other are also cut. Multi-site events are removed using the AvsE cut (Sec. 1.11). Finally, surface events with degraded energies due to a delayed charge component are cut using the Delayed Charge Recovery (DCR) cut (Sec. 1.12). The mean detection efficiency for single-site, bulk events, including $0\nu\beta\beta$, after applying these cuts is 0.803. The background spectrum before and after application of the AvsE and DCR cuts is shown in Fig. 1.9-3. The background index is calculated by counting all events that pass these cuts in the energy range from 1950 keV to 2350 keV. Regions including known γ-ray peaks and the $0\nu\beta\beta$ ROI are excluded. From this background index, a background of $4.0^{+3.1}_{-2.5}$ counts/(FWHM t yr) was measured, in the low background dataset excluding DS1 and DS5a. A limit on the rate of $0\nu\beta\beta$ is obtained using an unbinned extended profile likelihood fit with a flat background and using the previously mentioned measured peakshape function. The half-life limit, using the full 9.95 kg yr unblinded exposure of DS0-5b, is $T^{0\nu}_{1/2} > 1.9 \times 10^{25}$ yr, corresponding to an effective Majorana neutrino mass limit of $\langle m_{\beta\beta} \rangle < 240 - 520$ meV[1].

---

[1] C.E. Aalseth *et al.*, Phys. Rev. Lett. **120**, 132502 (2018).

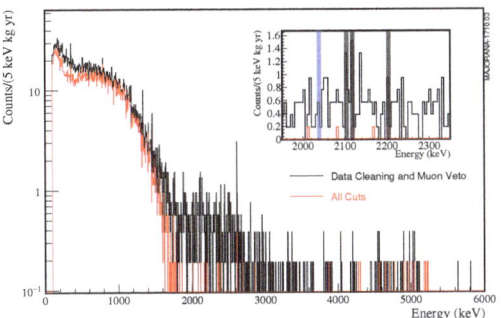 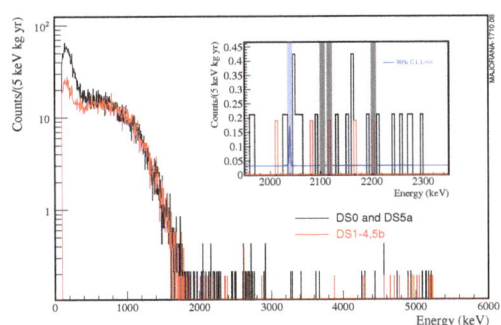

Figure 1.9-3. *Left:* Background spectrum of low background datasets (DS1-4, 5b). The black spectrum includes only data cleaning cuts applied before AvsE and DCR, while the red spectrum also includes AvsE and DCR. The inset contains the backgrounds in the region used to compute the background index. The shaded blue region is the $0\nu\beta\beta$ROI, and the shaded black regions are excluded due to known γ-peaks. *Right:* Comparison of high background datasets (DS0, 5a) to low background datasets.

1.10 Construction of a comprehensive model of radioactive backgrounds for the MAJORANA DEMONSTRATOR

S. Alvis, M. Buuck, J. A. Detwiler, Z. Fu*, J. Gruszko*, I. Guinn, W. Pettus, and N. Ruof

Work this year has been ongoing in creating the background model for the MAJORANA DEMONSTRATOR. A full mock-up of the experiment has been built in the GEANT4 simulation package[1], with the add-on MAGE[2], that specifies all of the experiment-specific geometry and particle tracking and generates ROOT[3] output files. GEANT4 is able to simulate the full dynamics of particle interactions with the various components of the DEMONSTRATOR model, including the shield, vacuum system parts, and the detectors themselves. The simulation records energy depositions in the detectors and saves those interactions to disk, where they are then additionally processed by GAT[4] to apply realistic temporal and spatial groupings, as well as the measured detector response function. See Fig. 1.10-1 and Fig. 1.10-2 for visualizations of the encoded geometry of the DEMONSTRATOR.

---

*Presently at Massachussetts Institute of Technology, Cambridge, MA.

[1] S. Agostinelli *et al.*, Nucl. Instrum. Methods A **506**, 250 (2003).
[2] M. Boswell *et al.*, Nuclear Science, IEEE Transactions on **58**, 1212 (2011).
[3] I. Antcheva *et al.*, Computer Physics Communications **180**, 2499 (2009).
[4] MAJORANA Collaboration, "Germanium Analysis Toolkit (GAT)." http://mjwiki.npl.washington.edu/bin/view/Majorana/GermaniumAnalysisToolkit. MAJORANA Collaboration internal document.

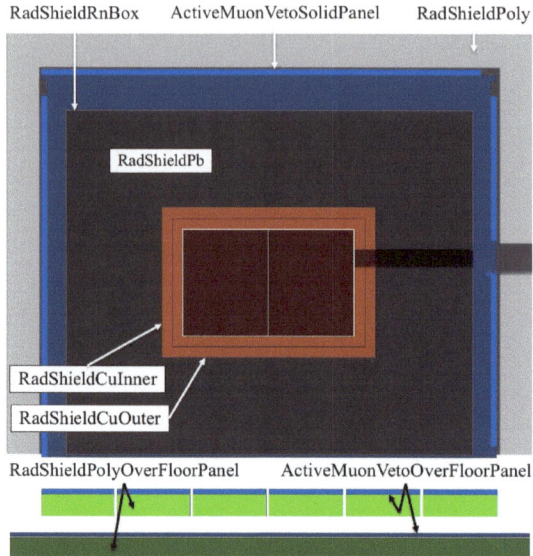

Figure 1.10-1. Cross-section of the radiation shield for the MAJORANA DEMONSTRATOR as constructed in the GEANT4 model.

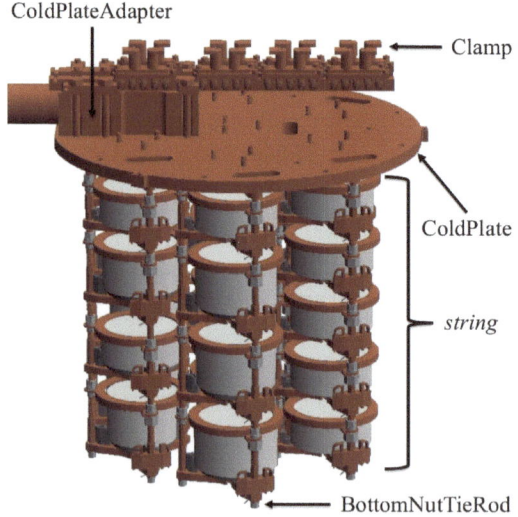

Figure 1.10-2. Strings of germanium detectors as they are arranged in the MAJORANA DEMONSTRATOR GEANT4 model.

The resulting simulated spectrum has been tested against the MAJORANA ^{228}Th calibration source, which produces a gamma line at 2614.5 keV with a detailed spectrum. The simulated spectrum matches the measured one with high fidelity, as shown in Fig. 1.10-3. The simulation is produced by simulating the decay chain of ^{228}Th to ^{208}Pb in the calibration source, tracking the emitted particles until they have deposited all of their kinetic energy or left the simulation volume, and applying just the Gaussian and low-energy tail components of the detector peak-shape function (article 1.9. The parameters of the peak-shape function that are not added in heuristically (the step height and the quadratic background) are approximated well by the simulation. However, the height of some of the peaks in the calibration data relative to the underlying Compton continuum appears to vary over time and therefore does not always fit well to the simulation. This is an active area of research to further refine the model.

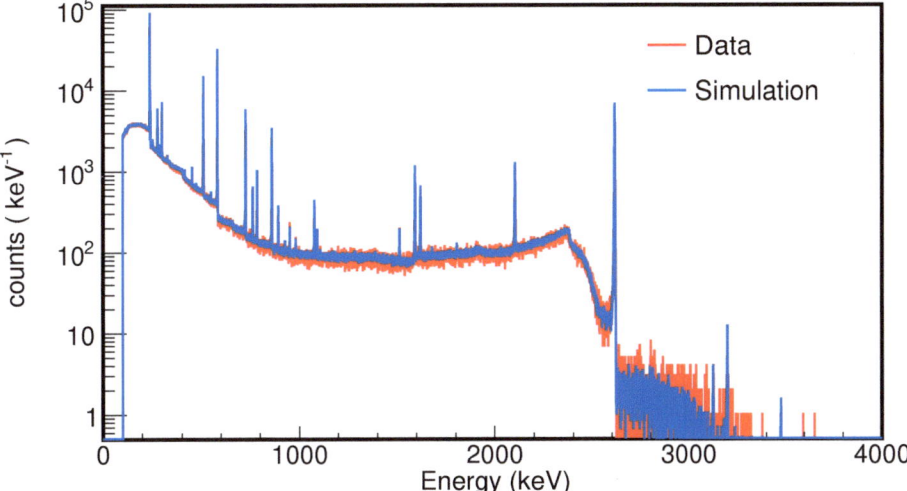

Figure 1.10-3. A comparison of the simulated calibration spectrum to the measured calibration spectrum. Simulation and experiment agree very well, except near the peaks of several of the calibration gamma lines.

In the same way that the calibration source is simulated, simulations of contaminants—such as ^{232}Th, ^{238}U, ^{40}K, ^{60}Co, and others—in different components of the experiment are simulated. We then construct the expected detection spectra normalized by the number of primary decays. These spectra are then multiplied by the expected activity in those components, based on either measured assay values or computed exposure/cosmogenic activation. This gives us an expected spectrum of counts/livetime, which can be converted to counts/exposure simply by dividing by the active mass of the detector. The expected background spectrum in the region near the ^{76}Ge $0\nu\beta\beta$-decay Q-value is shown in Fig. 1.10-4. The low-energy threshold is set at the point where the $2\nu\beta\beta$-decay spectrum becomes subdominant, and the high-energy threshold is set at the Compton edge of the 2615 keV gamma line from ^{208}Tl decay. The four notches correspond to known gamma lines from Bi and Tl, and the $0\nu\beta\beta$ decay window itself.

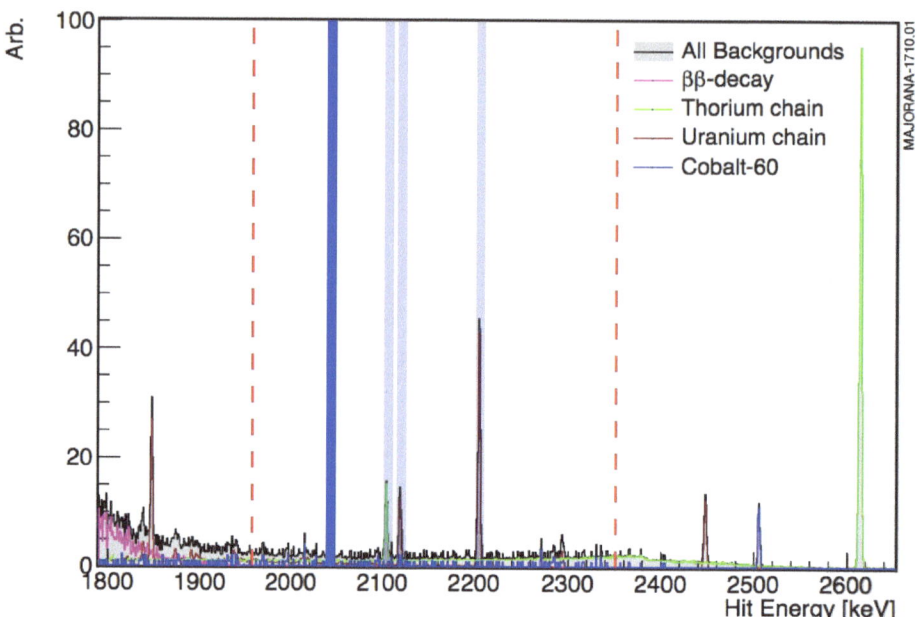

Figure 1.10-4. The expected spectrum near the Q-value for $0\nu\beta\beta$-decay in ^{76}Ge. The region we use to compute an average background for the Q-value is 1950–2034, 2044–2099, 2109–2113, 2123–2199, 2209–2350 keV, which excludes the $0\nu\beta\beta$ peak region as well as 3 known gamma peaks. This expected background in this region is approximately flat in energy, which is why it is used for the background estimation.

Efforts are ongoing to produce a version of this background model that is not tied to the expected contaminant activities, but rather fit to data taken with the DEMONSTRATOR. This will be an important step in the evaluation of the design and construction of the experiment, since a major goal of the experiment was to achieve an ultra-low background through careful design, materials selection and fabrication, and construction. We expect to have a working model fit to the data in the next few months.

1.11 Multi-site background rejection in the MAJORANA DEMONSTRATOR

S. I. Alvis, M. Buuck, C. Cuesta*, J. A. Detwiler, I. Guinn, W. Pettus, and N. W. Ruof

Among the key features of the P-type point contact (PPC) HPGe detector technology for low background experiments is its multi-site event discrimination. These detectors feature slow drift times and a localized weighting potential near the point contact, which give rise to distinct peaks in the current waveform for each energy deposition. Since MeV-scale betas ($Q_{\beta\beta} = 2039$ keV for ^{76}Ge) have a range in germanium of millimeters, while gammas of comparable energy will preferentially Compton scatter, betas can be discriminated from gammas

---

*Presently at Centro de Investigatciones Energéticas, Medio Ambientales Y Technológicas, Madrid, Spain.

based on their energy deposition site multiplicity. The maximum of the current waveform becomes an effective single-parameter discriminator of these event topologies Fig. 1.11-1. The corresponding cut based on this parameter developed by Dr. Clara Cuesta at UW[1] has an acceptance of 90% of signal events and a rejection of 94% of background events.

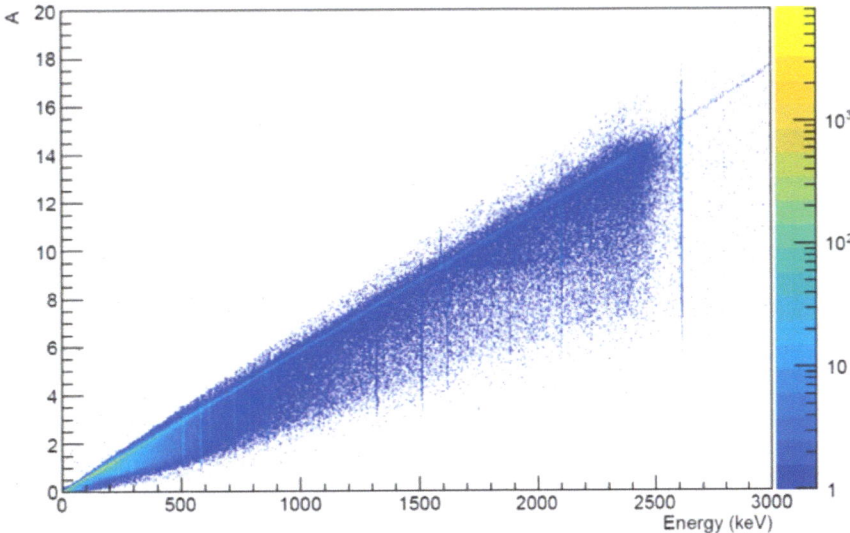

Figure 1.11-1. Distribution of maximum current (A) vs energy from ^{228}Th source calibration data. The maximum current increases approximately linear with energy, with the single site events concentrated near the top of the band. Gamma lines (*e.g.*, 2615 keV of ^{208}Tl) of the source are visible as vertical lines, with strong multi-site character cascading downward with suppressed maximum current.

The *AvsE* cut parameter used in analysis of the MAJORANA DEMONSTRATOR data had previously been well established, as well as its calibration procedure from ^{228}Th source calibration data. In preparation for the published result, the cut parameters for each dataset were reevaluated using the final energy calibration and channel selection. These parameters were validated to be consistent with previous analyses, but optimized for the final result.

Significant effort was made this year to better-quantify the systematic errors of the cut. This work was particularly important as the *AvsE* cut remains the driving contribution to the systematic error budget of the experiment. Improvements were made, particularly in assessing the stability of the cut performance and in understanding the shape of the underlying current distribution using simulated waveforms.

The *AvsE* cut is calibrated primarily from long (∼12 hr) calibration runs performed approximately once every few months. More regular short (∼1 hr) calibrations were also made to correct for gain shift of the detectors. Although low statistics prevent using the shorter runs to tune the cut parameters, they allow an aggregate analysis of the efficiency of the cut over time. Prof. Detwiler performed several statistical tests on these efficiency measurements and found the fluctuation of the data exceeded the expected statistical scatter

---

[1]CENPA Annual Report, University of Washington (2016) p. 17.

(see Fig. 1.11-2). The quoted stability systematic is conservatively corrected using the scatter in these data instead of the uncertainty on the average.

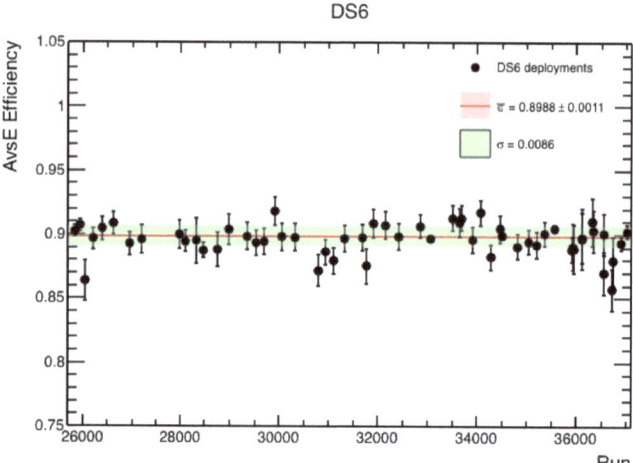

Figure 1.11-2. Stability of *AvsE* cut efficiency, as determined by the ^{208}Tl double-escape peak (DEP), over a yearlong period. The calculated average efficiency and error (red band, 1-σ) can be compared with the scatter in the data (green band). Despite being the longest dataset with consistent configuration, the achieved cut stability is among the best due to improved understanding of pulse shape instability associated with detector breakdown and rebiasing events.

The shape of the current-distribution may contribute to the cut systematics because the cut is tuned on double-escape peak (DEP) events at 1593 keV, but needs to function on neutrinoless double beta decay events at the $Q_{\beta\beta} = 2039$ keV. Because no sample of signal events at $Q_{\beta\beta}$ is available, these associated systematics were assessed using a full waveform simulation. In the future, this systematic will be crosschecked using a ^{56}Co calibration source currently on order, which has several DEPs above 2 MeV.

The waveform simulation is based on the `siggen` code that simulates charge collection in germanium detectors, combined with an analytic model of the waveform shaping from the DEMONSTRATOR electronics[1]. Events from the MAJORANA simulation (Sec. 1.10) are divided into local energy depositions and realistic waveforms are generated. These waveforms reproduced to high accuracy the features seen in measured data, and correspondingly mimicked the calibration current distributions. The ultimate limitation in this study was that electronics modeling is complete for only two detectors, so the systematic was conservatively assessed until more data are available. The systematic error was reduced by 40% relative to initial estimates, has been more robustly estimated (previously assumed based on data from another experiment with different detectors), and a mechanism is in place for improvement with additional data.

---

[1] B. Shanks, *Ph.D. thesis*, University of North Carolina (2017).

1.12 Characterizing surface alpha events for MAJORANA and LEGEND

S. Alvis, T. Bode[*†], M. Buuck, C. Cuesta[‡], J. A. Detwiler, J. Grusko[§], I. Guinn, S. Mertens[*†], and M. Willers[*]

Alpha background in the first $0\nu\beta\beta$ result

As one of our efforts to achieve a low background in MAJORANA DEMONSTRATOR, work continued on the DCR (delayed charge recovery) parameter. Delayed charge recovery is the phenomenon in which the charge released by an incoming particle is not released all at once. Instead, the last bit of charge is trapped and re-released on the timescale of waveform digitization. This occurs only on the passivated surface between the point contact and the dead layer of MAJORANA DEMONSTRATOR detectors, and is generally indicative of an external alpha particle. The quantified DCR parameter allows for an effective rejection of background alpha events while retaining $99 \pm 0.5\%$ of the bulk-detector events, as measured using ^{228}Th calibration source events.

Previously, MAJORANA and the Technical University of Munich collaborated to make the TUM (Technical University of Munich) Upside down BEGe (TUBE) scanner to study the effects of incoming alpha particles on the passivated surface of MAJORANA detectors [1]. Final results from the TUBE scanner show that the waveform response and energy are strong functions of the distance from the point contact to the location of the alpha event on the passivated surface. These results also indicate that the alpha activity is likely caused by ^{210}Po plated out as Rn daughters on Teflon components of the detector mount.

DCR tuning and stability in recent MJD data analysis

DCR is calculated using the tail slope from event waveforms. The tail slopes are caused by the electronics releasing charge to return to the baseline signal, and have consistent time constants. In calibration data, the slope of each waveform is a linear function of the energy, varying from detector to detector, with alpha events displaying higher slopes than calibration events at the same energy and detector. The DCR parameter is calculated by removing the linearity to make a horizontal spectrum with respect to energy, with a constant added so that 99% of the bulk detector events have a DCR parameter value of less than 0. The DCR data-cleaning cut is applied by requiring the DCR parameter value to be less than 0. Changes in detector bias and noise can tilt the tail slope vs energy distribution, and when this happens, the calculation of DCR needs to be redone, which is known as retuning. Retuning is done at the beginning of every dataset (on the order of months) with a 12-hour calibration run, to allow sufficient statistics to set a confident initial DCR cut. Additional retuning happens if significant noise is observed, or if rebiasing (or other instabilities) changes a detector's

---

[*]Technical University of Munich, Munich, Germany.
[†]Max Planck Institute, Munich, Germany.
[‡]Presently at Centro de Investigatciones Energéticas, Medio Ambientales Y Technológicas, Madrid, Spain.
[§]Presently at Massachusetts Institute of Technology, Boston, MA.

[1]M. Agostini, *et al.*, "BEGe detector response to alpha-induced energy depositions on the p+ electrode and groove surfaces." GERDA Scientific/Technical Report GSTR-13-006, Physik Department E15, Technische Universität München, July 2013.

energy spectrum significantly. This second retuning happens a 3-4 times per dataset, for 4-7 detectors.

Plotting the DCR cut efficiency over the duration of a dataset's calibration deployments allows us to inspect the DCR cut's efficiency (ideally 99%) and stability over time. This allows us to identify whether any anomalies exist and paves the way for more specific investigations if necessary. Though not all problems that appear in DCR can be fixed by retuning, the most common ones can be. At the present time, the retuning work is essentially complete, with few problems expected to arise in the most recent data.

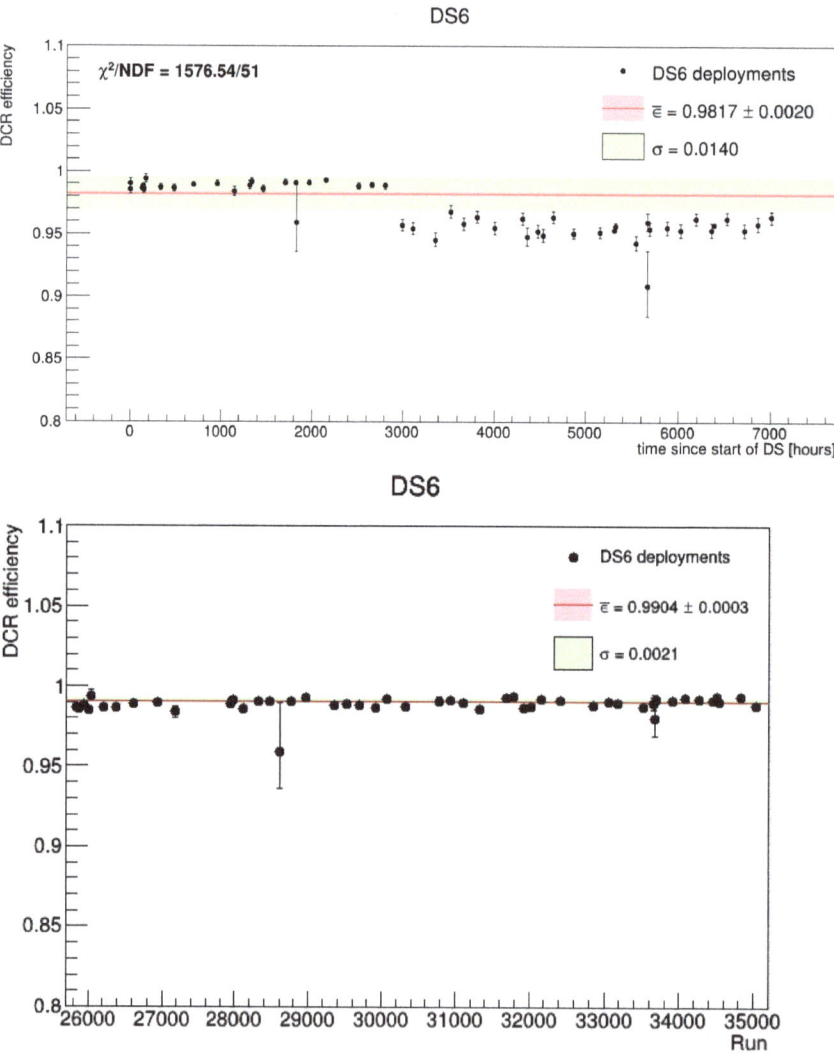

Figure 1.12-1. *Top:* Initial DCR stability study for DS6. This shows the efficiency of the DCR cut (ideally 99%) for all calibrations in DS6 (the past year). An average efficiency of 98.17% is significantly lower than desired, but is caused by the $\sim$3% lower efficiency in the second half of the dataset. Inspection of individual detector data yielded 3 problem detectors, which were subsequently retuned. *Bottom:* After retuning the problem detectors, the stability study shows an excellent overall efficiency of 99.04, consistent over the whole dataset.

Alpha response R&D for future Ge $0\nu\beta\beta$ and LEGEND

The DCR cut is set and studied using calibration data, but shifts in noise or biasing could happen at any time between calibration deployments. Since the most time-sensitive tasks were to set and tune the DCR cut, less attention has been given to which runs utilize which DCR retune, known as DCR run coverage. Currently, if there is a change in the DCR cut, it is applied to either the start of the calibration where changes are observed or the end of the previous calibration, whichever eliminates fewer events in data cleaning. This is a conservative choice, and in the past, has been estimated as a small uncertainty in our efficiencies, as very few runs are affected. The majority of the background data are blinded, so an accurate estimate of where to apply the cut is generally not possible.

Recently, we have shown that detector-injected pulser events can show shifts in the waveform slope that matches the shifts we see in calibration data. Since pulser signals are injected, non-physics events, we plan to use pulser events from background data to accurately shift run coverage in background data without violating our blinding scheme. Setting the run coverage this way will allow us to have a better reduction in alpha background for times near which there are changing conditions that necessitate a DCR retune.

Alpha response R&D for future Ge $0\nu\beta\beta$ and LEGEND

Our lab at UW has a Majorana-style HPGe detector, MJ60. We have plans to use it to help characterize alpha backgrounds for LEGEND. To improve working conditions with the detector, some physical upgrades have been made to the surrounding system. A liquid nitrogen-siphoning collar was designed, fabricated, installed, and tested to reduce the time we wait for the detector to heat up from weeks to days. A secondary, nested IR-blocking shield was designed, fabricated, and installed to prevent several IR shine paths. The bottom of the shield will also serve as a flat surface close to the detector where we can install a variable-position alpha source. Some work has been done and continues on improving the quality of electronic cabling to the detector, both for signal quality and to ease removal of the detector from the cryostat.

Once cabling improvements and fixes are complete, the detector will be tested to ensure our signal-quality problems have been fixed. Upon acquisition of an alpha source, we will test the detector's response. We also plan to acquire servos to move the alpha source without opening the cryostat in order to characterize the detector response as a function of source position.

1.13 Light readout with silicon photomultipliers for LEGEND

S. Alvis, M. Buuck, J. Detwiler, I. Guinn, D. A. Peterson, W. Pettus, N. W. Ruof, and T. D. Van Wechel

In LEGEND 200 and LEGEND 1000 (Sec. 1.8), enriched germanium detectors will be immersed in liquid argon to act as a scintillation background veto for uranium and thorium decay-chain products in the energy region around $Q_{\beta\beta}$. Photomultiplier tubes have been the conventional choice for measuring light read out at the single photon scale; however, due to recent developments in the past few years Silicon Photomultipliers (SiPMs) appear to be a more efficient alternative. SiPMs are small and can be easily tiled on surfaces, require operating voltages of less than 100 V, and are resistant to environmental extremes including high magnetic fields.

Prior to operation, SiPMs must be characterized by a set of performance parameters: Geiger discharge probability, cross-talk probability, afterpulsing probability, gain, and dark count rate. Equipment from the LArGe test stand[1] is being used as a dark box and cryogenic vessel for SiPM characterization. Progress has been made in measuring the dark-count rate and cross-talk probability from a Hamamatsu S12572-025P MPPC, as shown in Fig. 1.13-1.

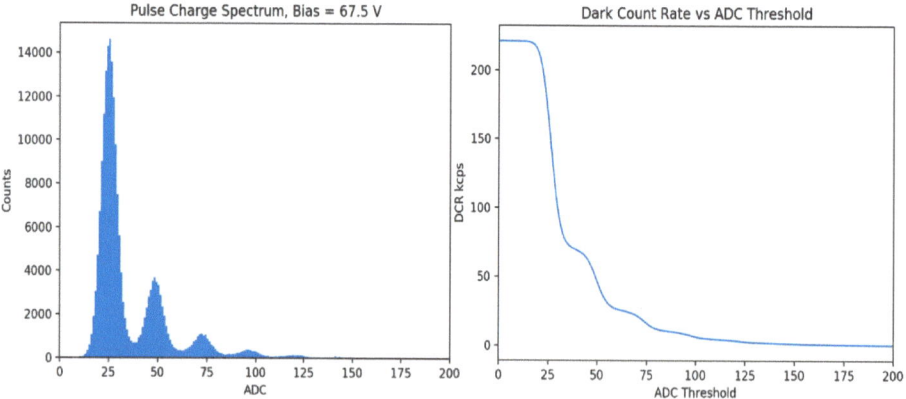

Figure 1.13-1. SiPM performance parameters such as the gain, dark count rate, and the cross-talk probability can be inferred from the analysis of the low-light pulse-charge spectrum (left) and a plot of the dark count rate vs threshold (right). SiPM pulses are integrated over a 100 ns time window to determine the amount of induced charge, which is histogrammed in the left plot. The peaks correspond to the charge produced by a single Geiger discharge, double Geiger discharge, etc. with the trigger-threshold placed halfway between zero and the single-discharge location. The cross-talk probability is measured by integrating the pulse-charge spectrum and producing the plot on the right. The ratio between the dark-count rate at the single-discharge and the double-discharge gives the cross-talk probability.

---

[1] CENPA Annual Report, University of Washington (2007) p. 27.

1.14 Simulations and analysis for LEGEND

S. I. Alvis, M. Buuck, J. A. Detwiler, I. Guinn, A. Hostiuc, W. Pettus, and N. W. Ruof

In this early stage of the experiment, LEGEND Simulations and Analysis tasks have focused on the development of a background model for LEGEND 200, and design of the software and computing infrastructure to support the analysis. UW personnel assisted in the development of LEGEND 200's first full background projections, prepared for the experiment's proposal to the LNGS Scientific Committee in early 2018. That effort built upon background modeling experience within MAJORANA, and made use of MaGe simulations with GAT post-processing tools. The UW effort focused primarily on modifications to the post-processing code to work with the newly-coded LEGEND 200 geometry.

Design of the LEGEND software and computing infrastructure is likewise also building upon MAJORANA experience, although we are investigating alternatives to the C++ and ROOT-based environment of GAT. We are investigating a file format based on HDF5, which has I/O libraries not just for C++ but also many other major languages. We are also investigating NumPy / SciPy as the primary analysis environment for typical users. We are evaluating prototype code in use at other institutions and plan to solidify the design over the coming months.

1.15 Forward-biased preamplifiers for LEGEND

S. I. Alvis, M. Buuck, J. A. Detwiler, I. Guinn, A. Hostiuc, D. A. Peterson, W. Pettus, N. W. Ruof, and T. D. Van Wechel

A charge sensitive preamplifier with forward biased reset is currently under development for the LEGEND project. The classic charge sensitive preamplifier circuit uses a feedback resistor for stabilizing the DC operating point and providing a DC path for the detector leakage current. The forward biased reset preamplifier eliminates the feedback resistor that is a major source of noise when using a low leakage detector. In LEGEND, this also eliminates the need to find a low-background resistor that can survive repeated LAr immersion, and thus represents an attractive alternative to the baseline technology. However, the DC path for the detector leakage current is through the JFET control gate for a forward biased charge preamplifier, so stabilization of the DC operating point must be provided by some other means.

Previously we had developed a forward biased reset preamplifier[1] that used dual gate tetrode JFETs manufactured by Moxtek for possible use by the MAJORANA or CoGeNT projects. The tetrode JFETs have two input gates: the detector anode is connected to the control gate, and a feedback capacitor connected between the output and the JFET control gate provides charge feedback. DC and low-frequency feedback to stabilize the DC operating point is applied by an RC network between the output and substrate gate.

---

[1] CENPA Annual Report, University of Washington (2015) p. 25.

Since the Moxtek tetrode JFETs are no longer available, the preamplifier currently under development is a forward biased charge sensitive preamplifier using the more commonly available triode JFETs that have a single input gate. Charge feedback is provided by a feedback capacitor connected between the output of the charge sensitive stage and the control gate as with any other charge sensitive amplifier. Stabilization of the DC operating point is provided by drain current feedback[1]. A feedback circuit regulates the JFET drain current to stabilize the DC operating point via DC/low-frequency feedback based upon the DC voltage at the output of the charge sensitive stage. The charge sensitive amplifier is split into two parts. The JFET and feedback capacitor are on a small front-end board at the detector containing only the JFET and the feedback capacitor for radio-purity reasons. The remainder of the preamplifier circuit is external to the cryostat. The JFET is followed by a folded cascode stage buffered by an emitter follower. A feedback capacitor to the JFET gate provides charge feedback. Stabilization of the DC operating point is provided by a discrete differential amplifier. The differential amplifier's input is connected to the output of the charge sensitive stage. Its output drives the top of the JFET's drain resistor. Therefore the JFET drain current is proportional to the DC voltage at the output of the differential amplifier stage. The charge sensitive stage is followed by voltage gain stages to provide low and high gain outputs.

The first prototype is currently under construction and will be tested this summer.

Project 8

1.16 Overview of Project 8

A. Ashtari Esfahani, R. Cervantes, P. J. Doe, M. Fertl, M. Guigue*, B. H. LaRoque*, E. Machado, E. Novitski, W. Pettus, R. G. H. Robertson, L. J. Rosenberg, and G. Rybka

Project 8 is a tritium-endpoint experimental program targeting neutrino-mass sensitivity ultimately covering the range allowed by the inverted hierarchy. The detector principle is the novel Cyclotron Radiation Emission Spectroscopy (CRES) technique, providing a differential energy spectrum measurement with high precision. The experiment has been conceived as a phased program spanning over a decade to achieve critical R&D milestones while simultaneously producing science results[2].

The collaboration is currently operating its "Phase II" apparatus. Following the first successful demonstration of the CRES technique[3], the apparatus was redesigned to increase the sensitive volume, optimize the signal-to-noise ratio, and allow use of tritium as a source gas. Operation with tritium will be the first measurement of a continuous spectrum using the

---

*Pacific Northwest National Laboratory, Richland, WA.

[1] F. Olschner and J.C. Lund, "Low Noise Charge Sensitive Preamplifier Using Drain Current Feedback", IEEE Conference on Nuclear Science Symposium and Medical Imaging, p. 378 (1992).
[2] A. Ashtari Esfahani *et al.*, J. Phys. G **44**, 054004 (2017).
[3] D. M. Asner *et al.*, Phys. Rev. Lett. **114**, 162501 (2015).

CRES technique, advancing from our previous measurements of the monoenergetic conversion electrons from ^{83m}Kr. The preparation for tritium operation involved a thorough safety review and extensive commissioning tests of the new gas system to ensure stable pressure regulation Fig. 1.16-1; these were completed in summer 2017.

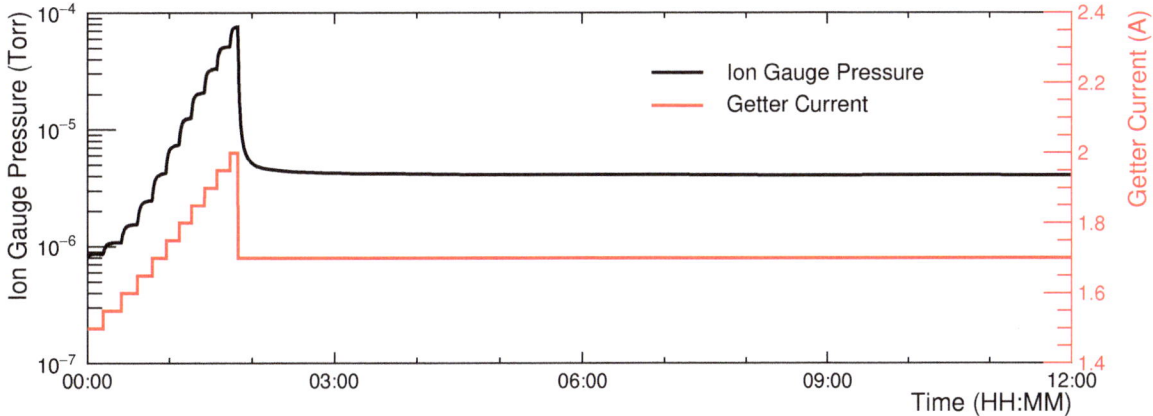

Figure 1.16-1. Tritium gas system pressure regulation with a getter during deuterium commissioning test. For the first two hours, the getter current is regularly increased in steps to demonstrate the tunability of the pressure setpoint. After 02:00, the getter current is held constant to demonstrate pressure stability.

As the host institution for Project 8, CENPA is home to personnel that continue to lead the hardware design and implementation, experiment operations, as well as DAQ and slow control development. A new undertaking last year was to perform an analytical derivation of the CRES signal features expected in waveguide. This work has already demonstrated its value in improving the resolution of the ^{83m}Kr line spectra (Sec. 1.17).

The next demonstration necessary for Project 8 will be to scale up the volume, and thus tritium decay statistics, which will be the focus of Phase III. The waveguide implementations of earlier phases are cm^3-scale or smaller, and do not present a clear path forward for significant scaling. Phase III will instrument a ~200 cm^3 volume, using either phased antenna arrays[1] or a multi-mode cavity (Sec. 1.18). In preparation for this next phase of the experiment, the B037 laboratory space has been newly reconfigured; the cleanroom has been removed to allow a restricted-access space encompassing the 5 G perimeter of the MRI magnet and the utility infrastructure is being planned.

Achieving the target neutrino mass sensitivity of 40 meV will require scaling Phase IV of the experiment to 10 m^3 of sensitive volume and switching to an atomic tritium source to avoid the final state broadening intrinsic to molecular tritium[2]. Overcoming the challenges inherent to production and trapping of atomic tritium is one more key focus area of the CENPA Project 8 group (Sec. 1.19).

---

[1] CENPA Annual Report, University of Washington (2017) p. 30.
[2] CENPA Annual Report, University of Washington (2017) p. 31.

1.17 Interpreting waveguide-effects on CRES signals

A. Ashtari Esfahani, R. Cervantes, P. J. Doe, M. Fertl, M. Guigue*, B. H. LaRoque*, E. Machado, E. Novitski, W. Pettus, R. G. H. Robertson, L. J. Rosenberg, and G. Rybka

The first two phases of Project 8 produce and trap energetic electrons inside a waveguide in the presence of a magnetic field. The waveguide constrains cyclotron radiation emitted by the electrons to propagate to a receiver with minimal losses. A reflector was included in the design to recapture the half of the power radiated away from the receiver and increase the SNR by 3 dBm. The magnetic field within the waveguide is a combination of a strong background field parallel with the waveguide axis and imposed distortions that form an electron-constraining magnetic bottle.

To characterize an electron's motion inside the trap, we define the pitch angle as the angle between the local magnetic field and the electron's velocity. As the electron moves toward regions of greater magnetic field its pitch angle will approach 90 degrees, whereas moving toward the decreased magnetic field at the bottom of the trap will cause its pitch angle to decrease. For initial pitch angles close-enough to 90 degree the trap will confine an electron. As they are reflected by each end of the magnetic bottle, these trapped electrons will make periodic axial oscillations inside the trap.

The electron's periodic motion introduces frequency modulation to the received signal via the Doppler effect. As a result, the signal power spectrum generated by an electron possesses a comb structure. The peaks in the spectrum will describe lines of detectable power, in a frequency-time spectrogram, which we call *tracks*. These tracks always have a positive slope, as each electron loses energy to radiation which gradually increases its cyclotron frequency.

The central track in a spectrogram is called the *main track* and its initial frequency is the measurement from which we extract the electron's initial energy. This frequency is directly related to the average magnetic field seen by the electron, implicitly a function of its initial pitch angle. As electrons with different initial pitch angles explore different regions of magnetic field, this effect will cause an apparent systematic offset in energy reconstruction if not properly characterized.

The axial frequency measurement is one way to find electron's pitch angle and account for the systematic. Spectrogram tracks parallel to the main track (*sidebands*), located at multiples of the axial frequency away from the main track, can be used to relate pitch angle and electron kinetic energy in the magnetic bottle traps (see Fig. 1.17-1).

---

*Pacific Northwest National Laboratory, Richland, WA.

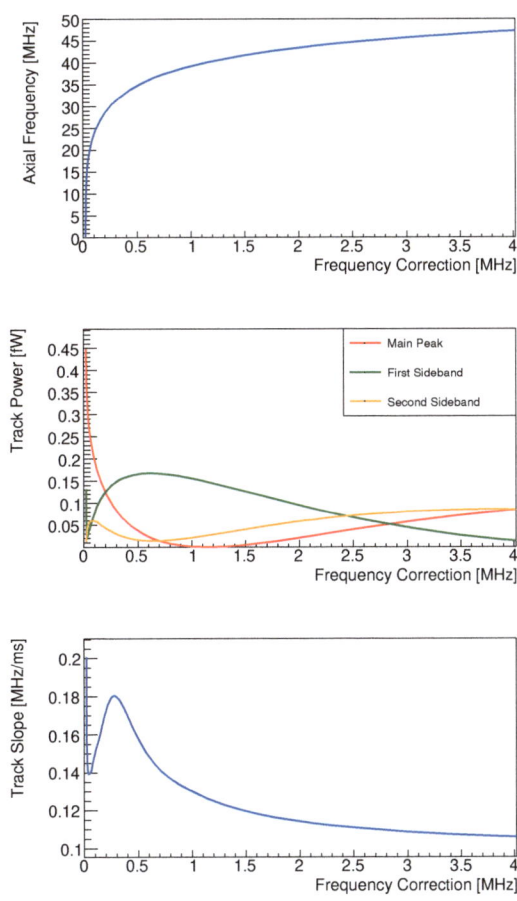

Figure 1.17-1. *(top)* Axial frequency, *()middle)* main peak and first two sidebands power, and *(bottom)* slope for 30 keV electrons vs. start frequency correction.

Track power offers another avenue for energy reconstruction. If the initial electron pitch angle decreases from 90 degrees, the doppler shift will increase and power will leak into sidebands (Fig. 1.17-1). However the cavity-interference caused by the waveguide's reflecting short can have a considerable effect on the observed power. If the main track happened to occur at a frequency at which a destructive interference is significant, the spectrogram can consist exclusively of sidebands. The 17-keV line of ^{83m}Kr was exactly placed in one of these nodes in the first phase of the experiment. A careful study of CRES signal features makes it possible to reconstruct this line's energy and restore the detector linearity (see Fig. 1.17-2).

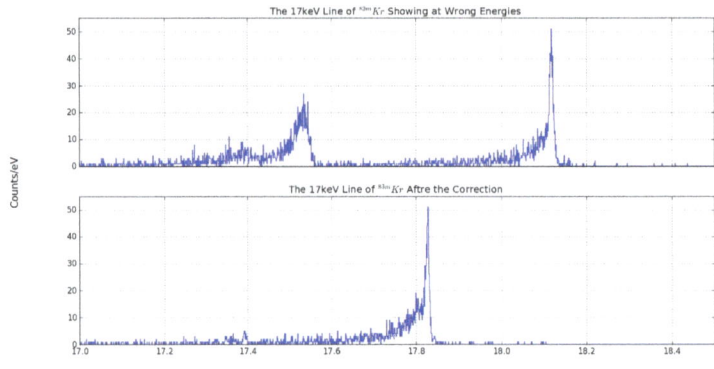

Figure 1.17-2. The 17 keV line before and after correction - it will change to root with Gaussian fit.

As mentioned earlier, track slope is a measure of electron's total radiated power. Therefore the information of radiated power can be extracted with a precise frequency measurement. In the presence of the waveguide reflector, the slope is one other way to find electron's pitch angle (Fig. 1.17-1).

1.18 A large-volume cavity CRES concept

A. Ashtari Esfahani, <u>R. Cervantes</u>, P. J. Doe, M. Fertl, M. Guigue*, B. H. LaRoque*, E. Machado, E. Novitski, W. Pettus, R. G. H. Robertson, L. J. Rosenberg, and G. Rybka

Project 8 is a tritium-endpoint experiment to measure the neutrino mass using the technique of Cyclotron Radiation Emission Spectroscopy (CRES). Early phases of Project 8 demonstrated CRES in small microwave waveguide volumes. An immediate challenge is to scale the detection scheme in size to accommodate sufficient tritium source intensity for a competitive neutrino mass measurement. This larger volume experiment, called Phase III, will demonstrate the receiver concept and its scalability to a much-larger final phase of Project 8, Phase IV.

Two design concepts are being explored in parallel for Phase III: a large free-space radiation environment and a resonant-cavity environment. Much progress has been made by collaborators from other institutions for the free space concept. At CENPA, we have worked on the multi-mode cavity concept. Once both concepts are mature, a decision will be made of which design is best suited for reconstructing CRES events in Phase III.

Significant progress has been made for the cavity concept, in both simulating CRES events and building a prototype to test whether events can be reconstructed. So far, we have simulated a stationary electron emitting 26 GHz dipole radiation in a cylindrical cavity 25 cm long and 12 cm in radius. We find that the electric field depends upon the position of the

---

*Pacific Northwest National Laboratory, Richland, WA.

electron (Fig. 1.18-1). We have also simulated the voltages induced in an antenna array inside of the cavity (Fig. 1.18-2). The detected voltages depend on the electron's location. From the induced voltages and phases, a radiating electron's position can be determined. Finally, we have successfully reconstructed the position of a radiating source in an experimental cavity using a likelihood fit (Fig. 1.18-3).

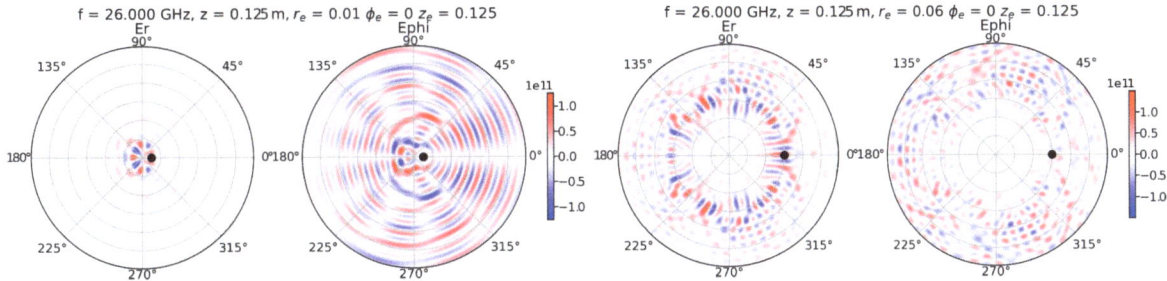

Figure 1.18-1. An electron emitting 26 GHz dipole radiation inside of a multimode cavity. The induced electric field depends on the position of the electron (represented as a black dot).

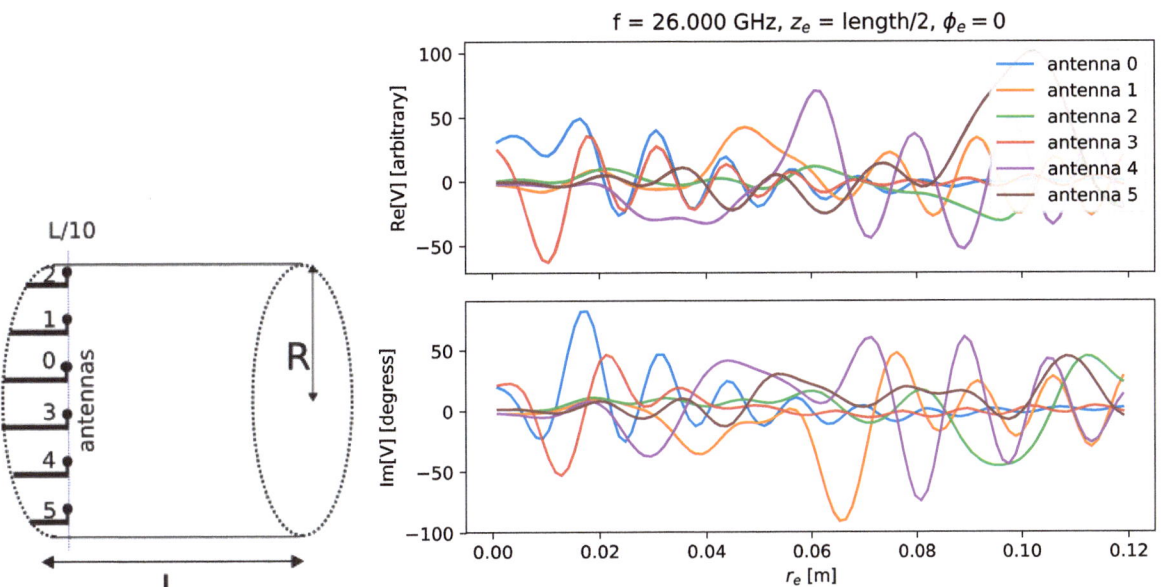

Figure 1.18-2. We move a simulated electron along the radius of the cavity and calculate the induced voltages on each antenna. Since the induced voltages change as a function of electron position, the voltages can be used to reconstruct the electron position.

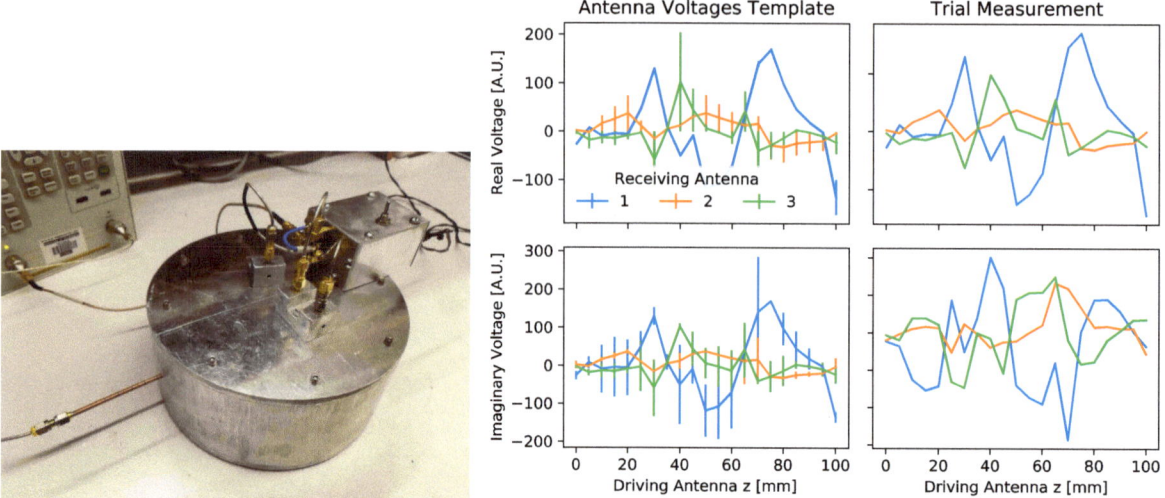

Figure 1.18-3. A cavity prototype that demonstrates the feasibility of determining the position of an emitting source. We measure the voltages of the three probing antennas as a function of the driving antenna position to form a fitting template. If the driving antenna is placed in an unknown position, we can perform a likelihood fit to successfully determine the source position.

1.19 R&D towards atomic tritium production for Project 8

A. Ashtari Esfahani, R. Cervantes, P. J. Doe, M. Fertl, M. Guigue*, B. H. LaRoque*, E. Machado, E. Novitski, W. Pettus, R. G. H. Robertson, L. J. Rosenberg, and G. Rybka

Throughout its phased approach, the Project 8 experiment is focused on demonstrating the technologies that will be required for a final 40 meV measurement of the electron neutrino mass, m_β. To meet that goal, molecular tritium cannot be used, as the highest energy portion of the beta spectrum is subject to distortions from the final state distribution of the molecule. Instead, a source of pure atomic tritium must be developed. To that end, research and development progress has been made at the CENPA, the Massachusetts Institute of Technology (MIT), Lawrence Livermore National Labs (LLNL), and the University of Mainz.

As in earlier phases of the experiment, the Cyclotron Radiation Emission Spectroscopy (CRES) detection volume is the same space in which the beta decay isotope is located. As CRES requires finite time to resolve a radiating electron, there is a practical lower bound upon the mean-free-path of electrons in the trap before a collision with an atom – the largest viable number density of atoms in the trap is 10^{12} atoms/cm^3. At that density, a roughly 10-m^3 volume of tritium atoms will be required to perform a CRES experiment that is sensitive to a neutrino mass of 40 meV after one year of data-taking. To complicate the matter further, the beta spectrum endpoint for molecular tritium is around 8 eV higher in energy than for atomic tritium, meaning that a purity of at least one part in 10^5 is required to meet the

---

*Pacific Northwest National Laboratory, Richland, WA.

target sensitivity. Tritium atoms recombine with one another to form molecules of T_2 in the event of three-body collisions. Interactions with surfaces can play the role of the third body for this recombination, meaning that any surfaces the atoms touch near the decay region will pollute the atomic tritium spectrum with molecular tritium decays. Fortunately, tritium atoms have a relatively large magnetic moment and experience a potential $U = -\vec{\mu} \cdot \vec{B}$ that allows them to be guided and trapped without touching physical surfaces.

The beam of atoms will start at a temperature $\sim$2500 K from a hot tungsten atomic "cracker". This class of crackers has several candidates for the final phase of the Project 8 experiment. The LLNL team is exploring a cracker that heats the tungsten capillary through high-voltage bombardment of electrons, while Mainz is exploring the use of a purely radiatively heated tungsten cracker. The CENPA team will explore the possibility of an inductively heated tungsten cracker that would function better in close proximity to large magnetic fields. Immediately following the cracker, the atoms will flow into an accommodator or set of accommodators, which are essentially tubes of a solid material held at a constant temperature, upon which the atoms thermalize. There is evidence to suggest that polished aluminum accommodators have a probability of recombination of 10^{-5} per bounce at temperatures of 140-160 K and are able to thermalize hydrogen and deuterium rapidly[1]. Eventually, such an accomodator would need to cool the atoms to 30 K to 60 K, and options for such an accommodator will be pursued in parallel at CENPA, LLNL, and Mainz. At CENPA, an option for immersing the accomodator in a 4-T to 5-T magnetic field and spin flipping high-field seeking atoms in order to velocity-select a larger fraction of the effusive Maxwell-Boltzmann velocity distribution will soon be explored.

The trapping region itself will require a 1 T uniform background field and steeply-rising 2-T high "walls" to contain atoms at a temperature of 30 mK with limited evaporative loss. The most promising trap configuration is a highly-multipolar Ioffe trap. Designs for such a trap have arisen from collaborators at Mainz and MIT[2]. The predominant trap losses are from atoms that leave the trap through the same hole in the field from which they entered. To limit these losses, a very small opening of roughly 1 cm^2 in the 2-T field contour is desired. Fig. 1.19-1 shows a preliminary design for this trap. Between the accommodator and the trap, atoms that are hotter than 700 mK must be selected and pumped away, so that the remaining atoms can climb the 1-T background field of the trap, losing kinetic energy to become 30 mK atoms. At CENPA, a magnetic thin lens for tritium atoms is being designed that will select 700 mK atoms and focus them to a tight spot so that they may enter the smallest possible hole in the magnetic trap, thus limiting losses from the trap as well as introducing the least possible contamination by hot atoms and molecules. Fig. 1.19-2 shows a simulation of the thin lens velocity selector.

---

[1]R. G. H. Robertson, et al, The Los Alamos experiment on atomic and molecular tritium, Los Alamos Nation Laboratory (1984).

[2]A. Radovinsky, Technical Note, Massachusetts Institute of Technology (2018).

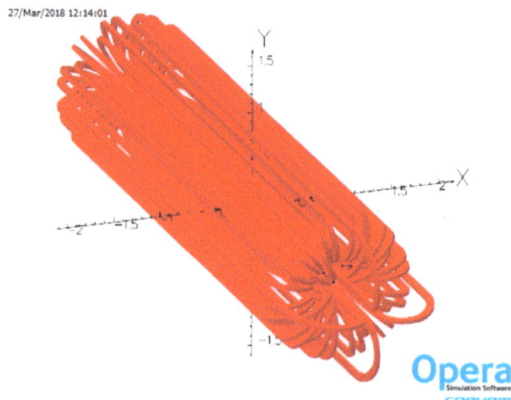

Figure 1.19-1. A preliminary design of the Ioffe trap for the final phase of Project 8

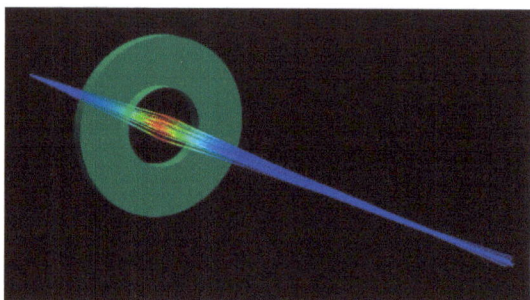

Figure 1.19-2. A Kassiopeia simulation of a velocity-selecting thin lens for tritium atoms

At present, the Project 8 experiment is the only direct neutrino mass experiment that focuses on using atomic tritium to escape the systematic effects from the final states of molecular tritium, meaning that extensive research and development of atomic hydrogen and tritium production, cooling, and trapping will be needed in the coming years. As Project 8 moves along with Phases II and III of the experiment, which explore many of the radiometric challenges of the experimental technique, there are many growing possibilities for parallel research and development at CENPA into the atomic tritium aspects of the final phase of the experiment[1].

---

[1] P. J. Doe, et al., Conceptual Design Report for Project 8: Measuring Neutrino Masses Using Frequency-Based Techniques, University of Washington (2017).

COHERENT

1.20 The COHERENT experiment: overview and first observation

J. A. Detwiler, E. M. Erkela, Z. Fu*, D. A. Peterson, <u>D. J. Salvat</u>, and T. D. Van Wechel

Coherent elastic neutrino-nucleus scattering (CEνNS) occurs when a neutrino scatters elastically from a nucleus with a momentum transfer that is small compared to the inverse size of the nucleus[1]. The neutrino thus scatters coherently from the constituent nucleons, and the resulting cross-section scales with the square of the number of neutrons in the nucleus. This greatly enhances the neutrino interaction rate compared to existing neutrino detection methods; at the same time, detecting the subsequent low energy nuclear recoils from the CEνNS interaction is challenging, and requires highly-sensitive detector technologies.

The resolution of these challenges provides a new means of studying neutrinos and nuclei. There is a robust prediction for CEνNS in the standard model, permitting tests of non-standard neutrino interactions[2], sterile neutrinos[3], and other neutrino properties[4]. CEνNS can potentially be used to monitor nuclear reactors[5], improve our understanding of core-collapse supernovae[6], improve the sensitivity of future weakly-interacting massive particle searches[7], and probe the structure of nuclei[8].

The Spallation Neutron Source (SNS) at ORNL is a unique venue for the development of CEvNS detection: it offers a high flux of pion decay-at-rest neutrinos, and its pulsed timing can be used to suppress backgrounds by ten-thousand-fold. The first observation of CEνNS was performed by the COHERENT collaboration using a 14.6 kg low-background CsI(Na) crystal[9]. A comprehensive study of fast neutron, neutrino-induced neutron, and environmental backgrounds guided a layered shielding configuration, and the detector was deployed 19.3 m from the SNS mercury spallation target. Background rejection, afterglow, and data quality cuts were developed using a series of radioactive source calibrations, including a forward compton-scatter-tagged ^{133}Ba source calibration. The expected neutrino flux, CsI quenching factor, and detector threshold were used to calculate the expected number of CEνNS events as a function of time relative to beam pulses and as a function of photo-electron yield. A maximum-likelihood fit to the observed signal (coincident with beam) and

---

*Presently at Massachusetts Institute of Technology, Boston, MA.

[1] D.Z. Freedman. *Phys. Rev.* **D9** 1389 (1974).
[2] J. Barranco, *et al. Phys. Rev.* **D76** 073008 (2007), P. deNiverville, *et al. Phys. Rev.* **D92** 095005 (2015), B. Dutta, *et al. Phys. Rev.* **D93** 013015 (2016).
[3] A.J. Anderson, *et al. Phys. Rev.* **D86** 013004 (2012), B. Dutta, *et al. Phys. Rev.* **D94** 093002 (2016).
[4] T.S. Kosmas, *et al. Phys. Rev.* **D92** 095005 (2015).
[5] Y. Kim. *Nucl. Eng. Tech.* **48** 285 (2016).
[6] J.R. Wilson. *Phys. Rev. Lett.* **32** 849 (1974), D.N. Schramm and W.D. Arnett. *Phys. Rev. Lett.* **34** 113 (1975), D.Z. Freedman, *et al. Ann. Rev. Nucl. Sci.* **27** 167 (1977).
[7] J. Bollard, *et al. Phys. Rev.* **D89** 023524 (2014).
[8] J.A. Formaggio and G.P. Zeller. *Rev. Mod. Phys.* **84** 1307 (2012).
[9] D. Akimov, *et al. Science* 10.1126/science.aao0990 (2017).

background (anti-coincident with beam) favors a non-zero CEνNS signal at 6.7 σ, in good agreement with the standard model prediction. The fitted signal and background versus time and photo-electron yield are shown in Fig. 1.20-1.

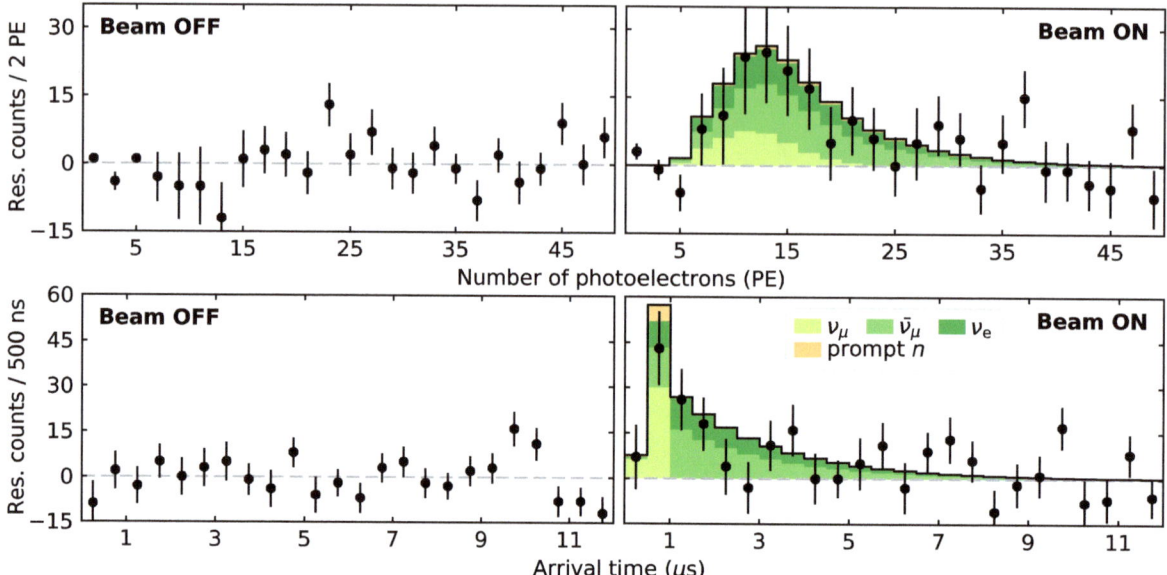

Figure 1.20-1. *Top:* The observed CEνNS signal above background (black dots) and expected signal and background contributions (colored bars) versus photo-electron yield. The "Beam OFF" data are used as a background model to subtract from the signal. The decrease in expected and observed signal below 5-10 PE is due to the photo-electron-dependent acceptance of events due to quality and pulse shape cuts. *Bottom:* The observed background and signal residuals as a function of time relative to protons on target. The prompt neutron signal is small compared to the signal, and the arrival time distributions show the characteristic prompt $\bar{\nu}_\mu$ contribution from π^+ decay, as well as the ν_μ and ν_e flux due to μ^+ decay.

The CsI(Na) detector will continue to take data to reduce statistical uncertainties, and there is an on-going effort to further reduce systematic effects related to the neutrino flux and quenching factor of nuclear recoils in CsI. Precisely testing the standard model prediction for CEνNS requires additional measurements with targets composed of both light and heavy nuclei. There are multiple current and planned detectors employed by the collaboration to perform a comprehensive study of CEνNS interactions, including a 10 kg fiducial-mass liquid argon scintillator detector (deployed), and a 10 kg array of P-type point contact germanium detectors. In addition, a 185 kg array of NaI(Tl) detectors (NaIνE) is currently in operation to detect the charged-current neutrino interaction ^{127}I$(\nu_e, e^-)^{127}$Xe* and serve as a prototype for a future tonne-scale NaI array[1].

---

[1] D. Akimov, *et al.* Arxiv:1803.09183.

1.21 Crystal characterization and low-cost PMT base development for a ton-scale NaI array for COHERENT

J. A. Detwiler, E. M. Erkela, Z. Fu, D. A. Peterson, D. J. Salvat, and T. D. Van Wechel

The NaI(Tl) crystal scintillators currently stored at CENPA are intended for incorporation into a multi-ton array at the SNS, with the primary goal of detecting low-energy neutral-current CEνNS events. This detector array will also perform a measurement of charged-current neutrino-scattering events. In order to enable simultaneous detection of these two types of events, which occur at energy regimes separated by roughly four orders of magnitude, a custom dual-gain PMT base is currently in development. To minimize cost, the bases are designed to run in a daisy-chain configuration using compact EMCO high-voltage units supplied with the original detectors received from the Department of Homeland Security. The base, pictured in Fig. 1.21-1, is in the final stages of development. We plan to produce enough bases to outfit the entire UW detector array.

Figure 1.21-1. One of the UW NaI(Tl) scintillating crystals resting in the crystal characterization platform and outfitted with a prototype of the low-cost PMT base under development at UW. The two black coaxial cables carry the low-gain and high-gain signals to the data-acquisition unit. The other connectors provide DC power and serial-data connections to program the EMCO high voltage units.

Prior to deploying the crystals, it is necessary to characterize their performance. From preliminary testing, the detector population at CENPA exhibits significant variation in baseline gain and voltage response. To correct for this, a characterization procedure was developed to map the resolution and gain characteristics of each crystal and to reject those which exhibit faults or other factors that would make them unreliable. Moreover, crystals with similar gain can be grouped together and powered by a single high voltage unit, reducing the cost of the final array. A crystal-characterization procedure was developed to measure gain curves and position variation for each crystal. A wooden platform with crystal and source holders (visible in Fig. 1.21-1) was constructed to ensure consistency and reproducibility for all measurements. A sophisticated spectral fitting routine was also developed to automatically calibrate and analyze the characterization data, allowing for variations across the scintillator population. A script automating the full analysis is nearly complete. An example gain curve output by the script is shown in Fig. 1.21-2. Full-scale characterization is expected to proceed in summer 2018.

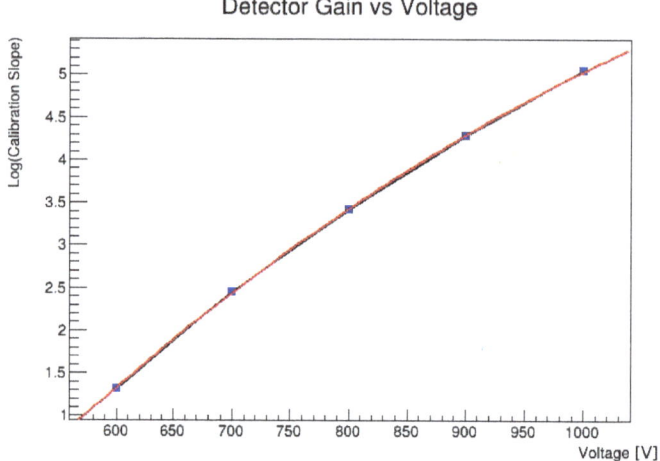

Figure 1.21-2. Detector gain versus voltage (blue dots) of a UW NaI(Tl) crystal for COHERENT as measured in our crystal characterization procedure, showing approximately-exponential behavior. The red curve is a quadratic fit used to characterize the voltage-dependence of the gain for each detector. This figure is automatically produced after acquiring data for a given crystal, and cataloged in an ELOG dedicated to crystal characterization.

1.22 The COHERENT experiment: NaI simulations

J. A. Detwiler, E. M. Erkela, and D. J. Salvat

The NaIvE detector is a 24 crystal, 185 kg array of NaI(Tl) detectors currently deployed at the Spallation Netutron Source. The goal of this deployment is to detect charged-current neutrino interactions in iodine, and guide the development of a ∼tonne scale array[1]. Background reduction is critical for a CEνNS search due to the feeble nature of neutrino interactions. This requires extensive modeling of potential detector geometries and radiation shielding designs for the experiment and validation of laboratory tests and existing data from the NaIvE array. A photograph of the NaIvE geometry and a rendering of the simulated geometry are shown in Fig. 1.22-1.

We use a general-purpose GEANT4 framework to simulate backgrounds in the future experiment, including intrinsic radioactivity such as thallium and potassium, cosmic ray muons, and fast neutrons. In addition, there is an exhaust pipe extending along the wall of the experimental hall which carries radioactive exhaust gas from the liquid mercury spallation target at the SNS. This generates a considerable flux of 511 keV γ-rays from e^+-e^- annihilation, and compton-scatter events from these γ-rays are a potential source of background. An preliminary shielding design consists of a 10 cm layer of Pb to shield these γ-rays, surrounding a

---

[1]D. Akimov, et al. arXiv:1803.09183.

 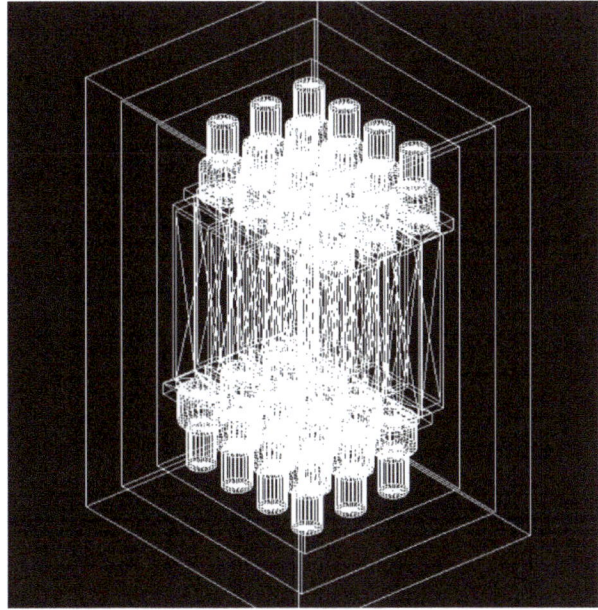

Figure 1.22-1. *Left:* The 185 kg NaIvE detector array during installation at the SNS. *Right:* The GEANT4 geometry of the 185 kg array with additional lead and water shielding (represented by the rectangular boxes around the detectors). This is a potential shielding configuration being investigated as part of the simulation effort.

10 cm layer of water to shield fast neutrons. We perform simulations of γ-rays incident upon the detector array with and without the shielding, and preliminary results indicate that the proposed shielding provides sufficient background reduction.

Further studies will focus on the shielding of fast neutrons, which produce low energy nuclear recoils indistinguishable from CEνNS. We will investigate water and polyethylene in various geometries, as well as potential neutron-absorbing material such as boron-loaded rubber. The geometry will be extended to include prototype designs for the tonne-scale array to aid in finalizing the design.

Selena

1.23 Selena R&D

<u>A. E. Chavarria</u> and A. Piers

We have recently proposed to develop imaging detectors with an active layer of amorphous Se (aSe) coupled to a complementary metal-oxide-semiconductor (CMOS) active pixel array[1]. Such devices will have the capability to image with high spatial and energy resolution the mm-long ionization tracks produced by minimum ionizing-electrons emitted in the $\beta\beta$ decay of ^{82}Se. We presented estimates on the expected backgrounds for a large imaging detector deployed in a low-radioactivity environment and showed that the proposed technology combines at once the possibility of a low-background implementation, the precise energy resolution required to reject background from the two-neutrino $\beta\beta$-decay channel, and the efficient determination of the event topology necessary for a powerful rejection of α, β and γ-ray backgrounds from natural radioactivity. Such a detector could be sensitive to $0\nu\beta\beta$ decay rates at the level of 10^{-6} decays/kg/year, opening the possibility to definitively test whether neutrinos are Majorana fermions in the case of a normal ordering of the three neutrino masses.

The excellent performance of the devices in terms of energy resolution is projected on the basis of the present literature. However, experimental data on the energy response of aSe are limited, with most of the research devoted to the understanding of the quality of radiographic images from $\sim$50 keV X-rays. To improve on these results and demonstrate the projected performance, we are building setups for X- and γ-ray spectroscopy with better electronics. Our first sensor was a 75 μm-thick aSe target layer sandwiched in between two electrodes, with a strong electric field (10–20 V/μm) applied across the aSe. A novel low-capacitance CMOS front end, the CUBE pulse-reset charge preamplifier[2], was connected to the cathode. We demonstrated a significant improvement in performance over previous aSe devices[3], with a baseline noise of the output signal of 35 e^- RMS, relative to the $\sim$1000 e^- expected from the absorption of 50 keV X-rays.

We exposed the aSe target to various X-ray radioactive sources and digitized the output waveforms with an analog-to-digital converter (ADC). The data is currently under analysis. Fig. 1.23-1 shows the waveforms from the absorption of 122 keV X-rays emitted by a ^{57}Co source. By visual inspection, we can already note crucial information that was not accessible for previous spectroscopy studies[3]. In particular, the dependence of the shape of the output pulses on the depth of the interaction in the aSe layer. In aSe, holes are 30 times more mobile than electrons, which has a significant impact on the response of the bipolar devices used for these studies, whose output signals have contributions from both charge carriers[4]. For example, when, on the one hand, the ionization produced by the X-ray absorption occurs at a short distance from the anode, the charge is collected promptly, as the slower carriers

---

[1] A. E. Chavarria, C. Galbiati, X. Li, and J. A. Rowlands, J. Instrum. **12**, P03022 (2017).
[2] P. Barton, M. Amman, R. Martin, and K. Vetter, Nucl. Inst. Meth. A **812**, 17 (2016).
[3] I. M. Blevis, D. C. Hunt, and J. A. Rowlands, J. Appl. Phys. **85**, 795B (1999).
[4] Z. He, Nucl. Inst. Meth. A **463**, 250 (2001).

(the electrons) need only travel a short distance to be collected, leading to a sharp rise time (Fig. 1.23-1 *top*). On the other hand, when the ionization occurs close the cathode, it takes longer for all the charge to be collected, as the electrons need to travel across the 75 μm of aSe, leading to much slower rise times (Fig. 1.23-1 *bottom*). Furthermore, the longer the electrons travel in the aSe, the more likely they are to fall into long-lived traps, which leads to a loss of signal.

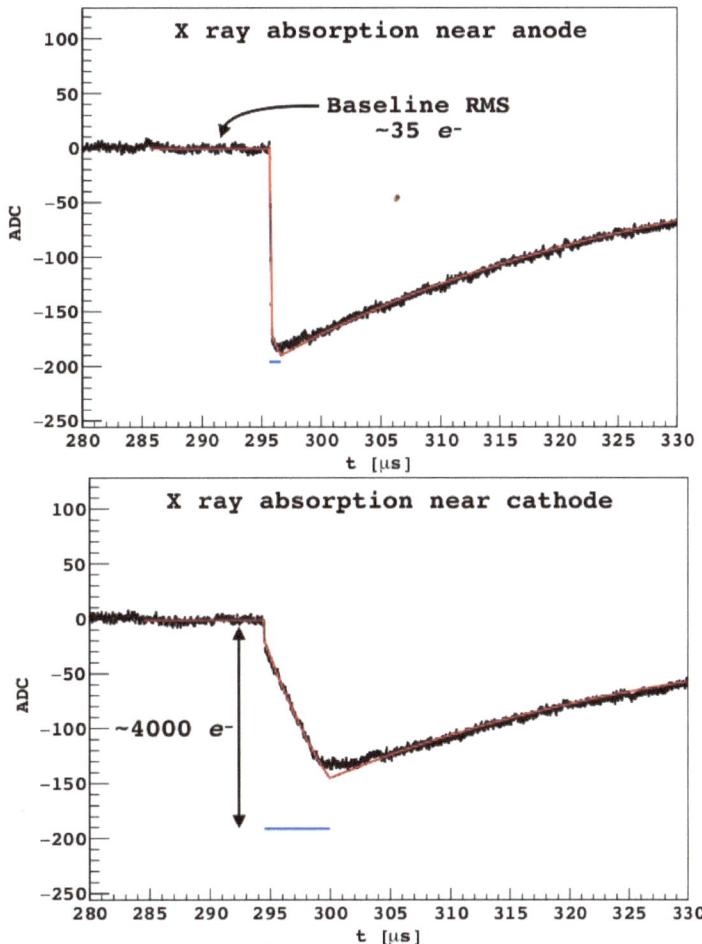

Figure 1.23-1. Digitized 122 keV X-ray pulses acquired with our low-noise setup. The rise time of the pulses is determined by the distance between the X-ray absorption and the collecting electrodes, while the decay time is set by the output band-pass noise filter. The horizontal blue line shows the corrected signal amplitude by a preliminary algorithm.

We are currently analyzing the time profiles of the pulses to understand the charge generation and transport properties of aSe and correct for these effects. However, the improvement in energy resolution that we can achieve by pulse-shape correction will be ultimately limited because the energy deposited by a single X-ray has a spatial distribution that extends throughout the active aSe layer, with signal pulses that are the superposition of ionizations at a wide range of distances from the collecting electrodes. Thus, similar pulse-shape parameters may arise from drastically different ionization patterns, hindering the energy reconstruction.

The negative effects in the energy response of aSe due to the slow mobility of the electrons can be strongly mitigated in unipolar charge sensors, which will lead to significant improvements in performance. The coplanar electrode design proposed by Luke[1] can be easily implemented in the devices we have been testing. In this scheme, the cathode plane is replaced with two interlaced electrodes held at slightly different potentials. Each electrode will be connected to an independent CUBE preamplifier circuit and the time-profiles of the signal pulses will be digitized. The difference in the induced current between the two electrodes is only sensitive to the holes as they drift in the vicinity of the cathode, effectively sensing only the hole signal. We have started the fabrication of the interlaced cathode structure on a glass substrate, on which aSe layered structures will be deposited. We are developing a full simulation of the unipolar devices, including the physics of X-ray absorption, the energy losses by the photoelectrons, the charge generation and transport in aSe, the induced current on the interlaced electrodes, and the response of the front-end electronics. Fig. 1.23-2 shows the potential in the aSe from the coplanar electrode structure modeled with COMSOL, which we use to optimize the geometry for best unipolar response. A unipolar device with a thicker 1 mm aSe layer will allow us to reliably calibrate the response to MeV-scale minimum-ionizing electrons, such as those emitted in $0\nu\beta\beta$.

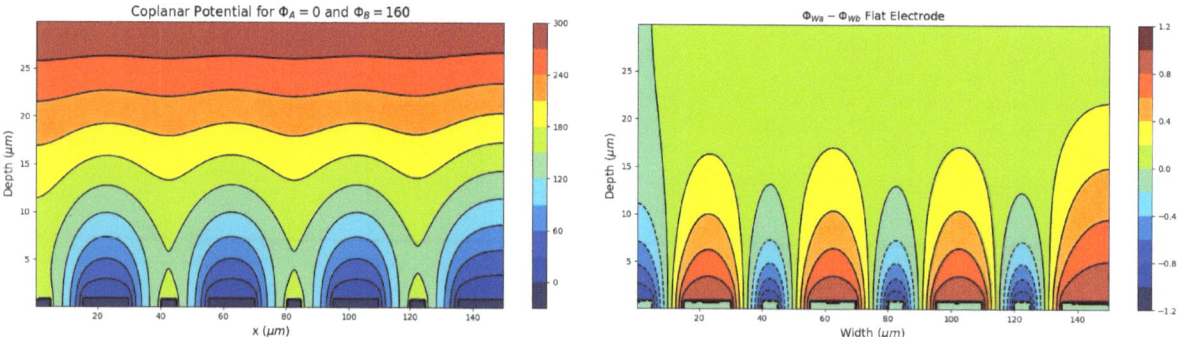

Figure 1.23-2. *Left:* Electrostatic potential in the aSe near the cathode plane. The geometry of, and potential difference between, the interlaced electrodes were chosen so that most of the field lines terminate on a single collection electrode. *Right:* "Weighing potential difference"[1] between the two interlaced electrodes, which quantifies the contribution to the output signal by a charge carrier as it drifts in the aSe. The fact that this value is almost zero except in the vicinity of the collection electrode signifies good unipolar response.

---

[1] P. N. Luke, Appl. Phys. Lett. **65**, 2884 (1994).

2 Non-accelerator-based tests of fundamental symmetries

Torsion-balance experiments

2.1 Fourier-Bessel gravitational inverse-square-law test

E. G. Adelberger, B. R. Heckel, J. G. Lee, and H. E. Swanson

New test-masses
A set of four new 50 μm-thick and two new 100 μm-thick platinum foils were cut in the wire-EDM at the physics shop. The thicknesses were all determined with the OGP SmartScope touch-probe before the cutting. The mass-loss in each cut was determined via repeated weighings and cleanings in isopropanol in the ultrasonic bath between cuttings. The pattern dimensions were determined with the SmartScope via edge-scans for each foil. Each foil was tested for possible magnetic contamination with a background subtracted GMR-probe turntable lock-in measurement. Four of the foils were epoxied with Stycast 1266 to glass annuli for use as test-masses in the experiment and gold-coated with the sputter source. Finally, a surface scan of 100 μm in radius and 50 μm in circumference spacing using the laser sensor of the SmartScope was performed on each gold-coated test-mass to evaluate surface roughness.

Closest separation to date
Two of the new test-masses were installed in the apparatus. After undergoing the standard tip-tilt alignments of the turntable, pendulum, screen, and apparatus, a separation of 62 μm was achieved – the closest separation to date of the platinum test-masses. Note that this separation is only on par with the closest measurements taken in 2012 by Ted Cook with Tungsten test-masses. Operating the pendulum at the closest separations is still limited by seismic activity that couples vertical motion to angle, from buses on nearby campus roads and construction equipment operating near CENPA.

Vertical position to angle coupling
We have made preliminary measurements of equilibrium angle as a function of pendulum to screen separation. There is a large dependence on the ability to align the test-mass of the pendulum to the screen.

First harmonic signal, 1ω
We found a spurious and nearly separation-independent signal at the attractor turntable rotation frequency, 1ω, that was about 10× larger than seen in Cook's thesis data. The 1ω signal was traced to a ~ 1 μrad flexing of the "Spider" relative to the lower portion of the outer vacuum chamber. This flexing is caused by a slight misalignment of the motor shaft attached to the outer vacuum can and the turntable axle suspended from the spider. The flexing was something seen in the older vacuum apparatus but was presumed to not be an issue with the new stiffer vacuum chamber and spider. The signal is largely cosmetic as it is unlikely that there is any contribution to the 18ω signal, let alone the 120ω signal. In addition, any contributions to the science signals can be measured with the pendulum

grossly miscentered from the turntable axis where the science signals should vanish. The introduction of a flexure bracing the spider against the outer vacuum chamber decreased the signal by ten-fold.

Ion Pump

An ion pump was added to the experiment to vibration isolation of the apparatus. This leaves the possibility of placing piezo stacks under the three feet that support the apparatus and implementing a vertical feedback control on a seismometer attached to the apparatus.

Dual LCR meter

A concept for simultaneously measuring the capacitance of the pendulum to electrostatic screen and attractor to electrostatic screen was developed with Zane Comden from Humboldt State. The device operated with a function generator providing a square-wave input to drive the electrostatic screen at approximately 1kHz, and four AD630 demodulator chips to lock-in both the in-phase and quadrature amplitude signals of the complex impedance of the two systems (both largely dominated by capacitance). The signals show little coupling to each other and are stable over multiple day measurements. The device must be calibrated using well-known capacitances.

The apparatus is currently open again to attempt to improve on the separation of the attractor test mass to the screen. We believe we can achieve a factor of 2 improvement. In addition we expect to begin implementing a test for the feasibility of the vertical isolation over the summer.

2.2 Progress on ground-rotation sensors for LIGO

J. H. Gundlach, C. A. Hagedorn, M. Ross, and K. Venkateswara

Over the last seven years, we have developed ultra-sensitive ground-rotation-sensors to improve seismic isolation in LIGO. The active seismic isolation system in Laser Gravitational-Wave Observatory (LIGO) utilizes ground seismometers to measure the ground translational motion and cancel it by actuating on the platforms from which the test masses are suspended in a feedforward loop. However, since seismometers are acceleration sensors, they are also sensitive to tilt coupling through gravity. During the first observation run in 2015 at the LIGO Hanford Observatory (LHO), this problem caused the low-frequency seismic isolation to perform poorly in wind speeds exceeding ~ 7 m/s, which reduced the duty cycle of gravitational-wave observation. Our rotation sensors, incorporated in the second observation run, were used to reduce tilt-noise in seismometers, thus making the LIGO interferometer capable of operating in wind speeds exceeding 15 m/s, hence improving the net duty cycle by $\sim 8\%$.

The rotation sensors consist of a low-frequency beam-balance in ultra-high vacuum, with an autocollimator for the angle readout. Ground horizontal acceleration coupling is rejected by locating the center of mass along the vertical very close to the suspension axis. The sensor can measure ground-rotations above 10 mHz with a sensitivity of 0.1 nrad/$\sqrt{(Hz)}$. Two of these sensors were constructed and installed at the two end-stations at LHO, before the start of O2, to subtract tilt-noise in collocated seismometers. At the vertex of the interferometer, one particular seismometer location was found to be less susceptible to wind-induced tilt since it was located far from the building walls. Thus using the two tiltmeters and the low-tilt seismometer location at the vertex, the robustness of the interferometer operation was improved in O2. Fig. 2.2-1 shows the fraction of time the LHO interferometer was locked as a function of wind-speed during O1 and O2. The improvement in duty cycle at speeds higher than 7 m/s is evident.

After successfully demonstrating their benefit to seismic isolation at LHO, the UW group was tasked with building similar sensors for LIGO Livingston Observatory (LLO). Due to smaller building floor space, no low-tilt location was found at the vertex at LLO. Therefore, four tiltmeters are being currently built and installed there, one close to each test mass. The parts were machined at the CENPA instrument shop and shipped to the site in February 2018 and work is expected to be completed by the end of May. With both LIGO sites made more robust against wind and weather conditions, more binary black hole and neutron star mergers will be measured during the third observation run, set to start close to the end of 2018.

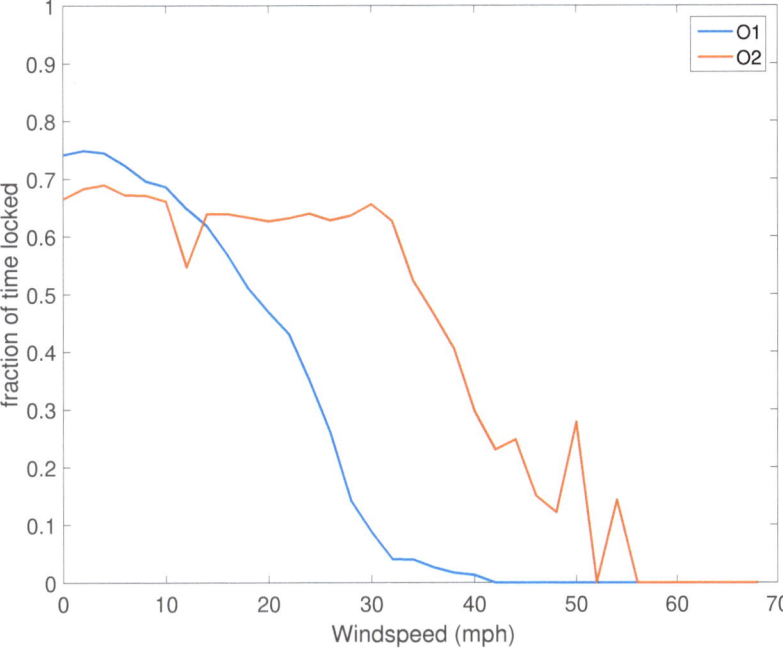

Figure 2.2-1. Plot of the fraction of time LHO was able to make gravitational-wave observations as a function of wind-speed during O1 and O2. The improvement in duty cycle at speeds higher than 7 m/s was enabled by the use of our ground rotation sensors.

2.3 An interferometric torsion-balance for atmospheric newtonian noise measurements

J. H. Gundlach, C. A. Hagedorn, M. Ross, and K. Venkateswara

We have initiated an experiment to test an interferometric angular readout on a torsion balance to achieve $\sim$picorad/$\sqrt{\text{(Hz)}}$ angular sensitivity. Such a readout could be used to improve multiple existing torsion balance experiments. Also, if combined with a pendulum with a large mass quadrupole, it could be used to measure the gravitational forces from atmospheric density fluctuations. This atmospheric Newtonian noise is expected to be a fundamental noise source for third generation gravitational-wave (GW) detectors. While it can't be measured directly at the frequencies relevant to GW detectors, models for this type of noise could be tested at lower frequencies where their effects are much larger.

The experiment consists of a dumbbell-shaped torsion balance with a dual angle readout - an interferometric readout for small dynamic range but high sensitivity, and a more traditional autocollimator readout for larger range operation. The interferometric readout is a modified Michelson interferometer with two steering mirrors to direct the beams onto the mirrors at the ends of the torsion balance which then recombine at the beam-splitter to form an interference pattern at the two photo-diodes (Fig. 2.3-1). Since the interferometric readout is linear only in a small range of motion near the center, the pendulum is held in feedback using two capacitive actuators. Fig. 2.3-2 shows a photograph of the setup with the interferometer parts in the foreground and the gold-coated torsion pendulum and capacitor plates in the background.

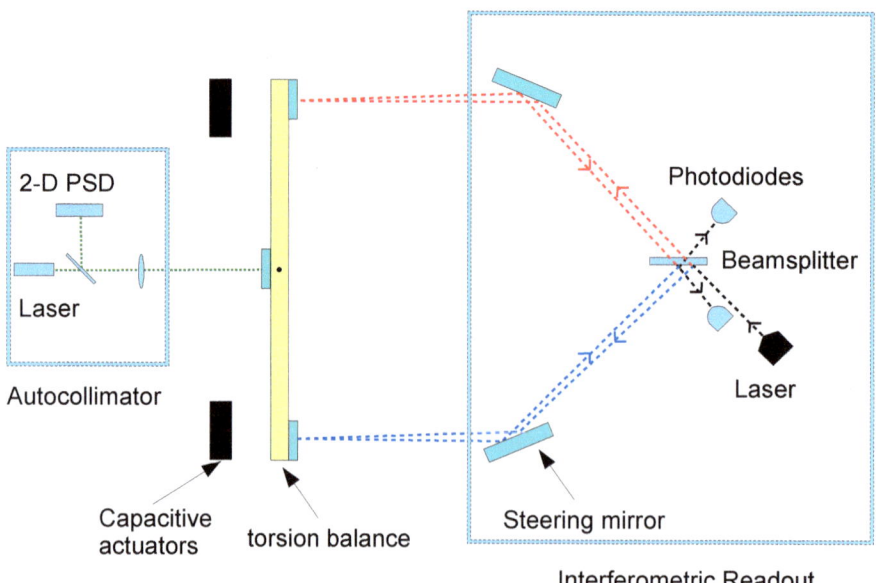

Figure 2.3-1. Schematic of the top view of the current setup. The torsion balance (yellow beam) has a large mass quadrupole and the torsion fiber extends out of plane.

Figure 2.3-2. Picture of the current setup.

Currently, the angular readout was dominated by the fiber thermal noise at low frequencies below 10 mHz. At frequencies higher than 0.1 Hz, the readout achieves about 0.1 nanorad/$\sqrt{(\text{Hz})}$, which appears to be limited by noise from scattered light. At intermediate frequencies, the noise seems to be from seismic origin. To reduce this noise, we plan to install a second identical torsion balance to make a differential angle measurement and reduce the common-mode seismic noise.

2.4 Preliminary limits on B-L coupled ultralight dark matter

E. G. Adelberger J. H. Gundlach, C. A. Hagedorn, J. G. Lee, E. A. Shaw, and K. Venkateswara

A recent theoretical proposal[1] motivates experiments for setting limits on the mass and coupling constants of a Baryon minus Lepton number (B-L) coupled dark matter candidate with $m_{DM} \ll 0.1$ eV. B-L is a conserved quantity in all known interactions and remains so in grand-unified and supersymmetric theories beyond the Standard Model. This makes it a well motivated conserved charge that could have a corresponding massive gauge boson coupled to it. We can search for a torque signal from this dark matter candidate with a B-L composition dipole pendulum such as the 8-test-body pendulum. The signal from this field-like dark matter would be oscillating at the dark matter Compton frequency with a coherent amplitude in an unknown direction in the frame of our solar system. As such, it would be modulated on terrestrial experiments at the sidereal day. Setting new limits requires improvements in thermal noise in the torsion fiber and the angular readout. Additionally, a rotating experiment allows us to modulate the signal at a frequency where the torsion balance is optimally sensitive.

In the past few years we have developed a fabrication method for fused silica fibers up to a length of 1 meter and demonstrated quality factors of up to $Q = 500,000$ in the LISA

---

[1] P. W. Graham, D. E. Kaplan, J. Mardon, S Rajendran, and W. A. Terrano, Phys. Rev. D **93**, 075029 (2016).

apparatus[1,2]. Compared to the quality factors of tungsten fibers with $Q \sim 5000$, the fused silica fibers in the Dark-EP experiment have lower thermal torque noise, $\delta\tau(f) = \sqrt{2k_B T \kappa/(\pi Q f)}$, where k_B is Boltzman's constant, T the fiber temperature, κ the fiber torsion constant, and Q the mechanical quality factor [3]. At room temperature the salient quantity to determine the relative sensitivity of an apparatus is $\sqrt{\kappa/Q}$. For tungsten fibers $\sqrt{\kappa/Q} = \sqrt{2.5 \times 10^{-9}/5000}$ $\sqrt{\text{Nm}}$ and we achieved $\sqrt{8.1 \times 10^{-9}/500000}\sqrt{\text{Nm}}$ with a fused silica fiber and our 8-test-body pendulum (Fig. 2.4-1)[1,2,4]. This corresponds to about a factor 5 improvement in torque sensitivity. A preliminary analysis of months of data demonstrates that we can take advantage of this improved sensitivity to set limits on this dark matter candidate over a longer timespan than has been shown in the past (Fig. 2.4-2)[5].

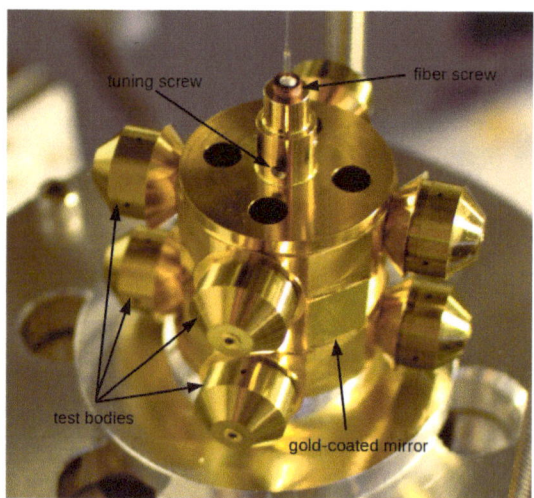

Figure 2.4-1. The 8-test-body pendulum. Currently it has 4-Al test bodies on one end and 4-Be test bodies on the other, which can be swapped with a set of Ti test bodies. This test mass distribution forms the B-L dipole for the experiment. It is suspended by a fused silica fiber in the LISA apparatus. The fiber is pulled from 1mm diameter rod stock glued into a fiber screw that screws into the pendulum body. In this experiment we use an improved autocollimator design, which measures the angular position of one of the pendulum's 4 gold-coated mirrors. Another feature is a set of 4-40 screws that tune the q_{21} gravitation moment of the pendulum. These mitigate gravity gradient systematics in our rotating equivalence principle experiment.

---

[1] CENPA Annual Report, University of Washington (2015) p. 30.
[2] CENPA Annual Report, University of Washington (2016) p. 34.
[3] C. A. Hagedorn, S. Schlamminger, and J. H. Gundlach, 2006 Proc. 6th Int. LISA Symp. (AIP Conf.Proc. vol 873) (New York, 2006) pp 189-193.
[4] CENPA Annual Report, University of Washington (2013) p. 47.
[5] W. H. Press and G. B. Rybicki, The Astrophysical Journal **338**, 277 (1989).

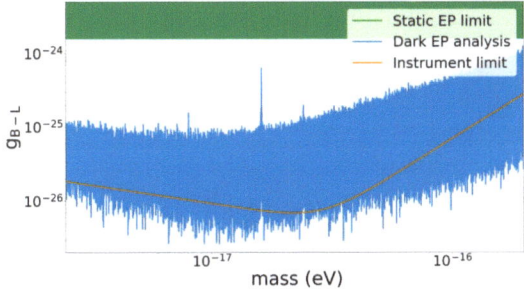

Figure 2.4-2. The Static EP limits are the current best limits set on an interaction, mediated by a B-L coupled vector boson, between the mass distribution of the earth and the 8-test-body pendulum in our rotating equivalence principle experiment. It is an indirect limit on this dark matter candidate. The Dark EP limit is a new preliminary limit on a direct interaction between this dark matter candidate and the 8-test-body pendulum. The spectral analysis is done using an approximate method known as the fast Lomb-Scargle periodogram that is useful for characterizing long data sets. This was performed on a dataset with 2,458 hrs of data. The visible peaks are known to be caused by nonlinearities in the angular readout. The instrument limit is the theoretical noise limit of the instrument, which assumes that the fiber has $Q = 500,000$, the pendulum has a moment of inertia of 3.78×10^{-5} kg/m^2, and the angular readout has a noise floor of 1 nrad/$\sqrt{\text{Hz}}$.

Other tests of fundamental symmetries

2.5 The mercury electric-dipole-moment experiment

Y. Chen*, B. Graner, B. Heckel, M. Ivory, and E. Lindahl[†]

Atomic electric dipole moment (EDM) measurements offer a sensitive probe of CP violation in theories of physics beyond the Standard Model. Our measurements of the ^{199}Hg EDM over the years have achieved the most precise EDM limit of any system. Searches in diamagnetic atoms, such as ^{199}Hg, provide bounds on nuclear EDMs that would most likely originate from CP violating interactions within and between nucleons. Our current upper limit for the EDM of ^{199}Hg is: $|d(^{199}\text{Hg})| < 7.4 \times 10^{-30}$ e · cm, a five-fold improvement in the upper limit over our previous work. [1]

The Hg EDM experiment is an NSF project whose grant was renewed in 2017 for another 4 year cycle (NSF Grant 1707573, P.I. Heckel). In 2017, Brent Graner (who was supported by CENPA) successfully defended his dissertation and has accepted a postdoctoral position at CENPA. Jennie Chen (supported by the NSF) is writing her thesis after re-analyzing the EDM data to search for a sinusoidal time-varying edm signal due to hypothesized axion-like particles. In 2018, Megan Ivory joined the project as a 50% FTE postdoc (supported by CENPA).

---

*University of Washington graduate student supported by NSF.
[†]University of Washington glass blower.

[1]B. Graner, Y. Chen, E. G. Lindahl, and B. R. Heckel, Phys. Rev. Lett. **116**, 161601 (2016).

A limiting factor in our recent EDM measurement was magnetic field gradients at the level of 0.2 µG/cm. The field gradients caused the Hg spin precession frequencies to depend upon how the Hg atoms averaged the vapor cell volume, leading to excess frequency noise from cycle to cycle. In addition, the field gradients led to a new systematic error due to small motions of the vapor cells from forces exerted by the applied electric field. To mitigate both problems, the magnetic field gradients must be reduced. The magnetic field is produced by an open-ended cosine theta coil with correction coils to account for the open ends. Using analytical techniques developed by Christopher Crawford at the University of Kentucky, we were able to design new correction coils that better satisfy the boundary conditions imposed by the magnetic shield that surrounds the cosine theta coil[1]. The new correction coils are being fabricated in our instrument shop, and magnetic field maps will commence in the summer of 2018. We anticipate a factor of 5 reduction in the magnetic field gradients.

We have also undertaken studies of our de-Gauss procedures, as we found that the field gradients would change after de-Gaussing. It was found that the transformer used to remove the DC component of the de-Gauss current was introducing harmonics of the exponentially decaying sine-wave current, leading to irreproducible de-Gauss cycles. We have rebuilt the de-Gauss electronics to remove the harmonics and have installed a second set of de-Gauss wires so that the inner and outer magnetic shields can be de-Gaussed independently. With a more uniform magnetic field and a more reproducible de-Gauss protocol, we believe the EDM apparatus will be ready for a new measurement cycle.

---

[1] https://indico.psi.ch/contributionDisplay.py?contribId=45&sessionId=22&confId=2973

3 Accelerator-based physics

3.1 Angular distribution of $2_1^+ \to 3_1^+$ photons in $^{21}\text{Ne}(p,\gamma)^{22}\text{Na}$ towards solving a puzzle

G. C. Ball*, R. Dunlop[†], A. García, A. Garnsworthy*, P. E. Garrett[†], D. S. Jamieson[†], J. Pedersen, B. Ribeiro[‡], E. Smith, M. Stortini, M. Sung, C. E. Svensson[†], and S. Triambak[‡§]

A measurement of the $\beta-\gamma$ directional coefficient in ^{22}Na beta decay has been used to extract recoil-order form factors[1]. The data indicated the requirement of a significantly large induced-tensor matrix element for the decay, well beyond the calculated first-class contribution. This suggests either second-class currents, which would disagree with other measurements, or a problem with some of the data that were used to arrive at the conclusion.

The conclusion relies heavily on the weak-magnetism form factor for the decay which was determined using an unpublished value of the analog $2_1^+ \to 3_1^+$ γ branch in ^{22}Na with the further assumption that the transition was purely iso-vector M1. Fig. 3.1-1 shows the decay scheme.

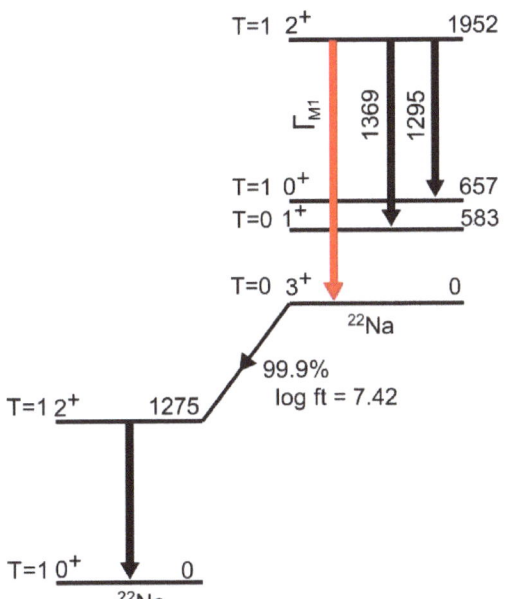

Figure 3.1-1. Decay scheme showing the 2_1^+ state. The branch in question in question is highlighted in red.

---

*TRIUMF, 4004 Wesbrook Mall, Vancouver, British Columbia, Canada.
[†]Department of Physics, University of Guelph, Guelph, Ontario, Canada.
[‡]Department of Physics and Astronomy, University of the Western Cape, South Africa.
[§]iThemba LABS, Somerset West, South Africa.

[1]C. J. Bowers et al., Phys. Rev. C **59**, 1113 (1999).

We produced implanted ^{21}Ne targets and used the accelerator in terminal-ion-source mode to feed the 2_1^+ state in ^{22}Na using a p, γ resonance. Our results on the branch have been recently published[1]. We obtained for the first time an unambiguous determination of the $2_1^+ \to 3_1^+$ branch in ^{22}Na to be 0.45(8)% compared to the previous unpublished value used before of 0.61(24)%.

To extract the weak-magnetism form factor from the measured branch, which one needs to extract the second-class-current form factor, one must identify the character of the $2_1^+ \to 3_1^+$ transition: the weak-magnetism is related only to the M1 width. It turns out that ^{22}Na is a deformed nucleus and it is likely that the transition is E2 dominated. An angular distribution could determine this clearly. Over the past year we have performed simulations which indicate that the alignment from the same reaction used before should allow for a determination of the mixing ratio to the level required. We have also revived an effort to produce new targets for the upcoming experiment. For the experiment, the tandem needs to be configured in its terminal ion source mode. We expect to run by the end of 2018.

3.2 Overview of the ^{6}He experiments

Y. S. Bagdasarova, K. Bailey*, M. Fertl, X. Fléchard[†], A. García, B. Graner,
M. Guigue[‡], R. Hong*, K. S. Khaw, A. Knecht[¶], A. Leredde*, P. Müller*,
O. Naviliat-Cuncic[||], T. P. O'Connor*, N. S. Oblath[‡], J. Pedersen, R. G. H. Robertson,
G. Rybka, G. Savard*, E. Smith, D. Stancil** D. W. Storm, H. E. Swanson,
B. A. VanDevender[‡], and A. R. Young[‡‡]

We are searching for tensor currents in two different experiments looking at the beta decay of ^{6}He. In an on-going effort we use laser traps to determine the $\beta - \nu$ correlation parameter a. A second effort, to determine the shape of the *beta* spectrum using the CRES technique, is being designed.

Tensor currents flip chirality and are forbidden in the standard model but could appear in precision beta decay experiments as evidence for physics at energies beyond the electroweak scale[2]. Fig. 3.2-1 shows limits from several sources. The limits from the LHC obtained by Cirigliano et al. are presently similar to those from beta decays[3], but will improve as the LHC gathers more data. The band shown in Fig. 3.2-1 shows limits expected after the LHC

---

*Physics Division, Argonne National Laboratory, Argonne, IL.
[†]LPC-Caen, ENSICAEN, Université de Caen, Caen, France.
[‡]Pacific Northwest National Laboratory, Richland, WA.
[¶]Paul Scherrer Institute, Villigen, Switzerland.
[||]Department of Physics and Astronomy and National Superconducting Cyclotron Laboratory, Michigan State University, East Lansing, MI.
**Electrical Engineering Department, North Carolina State University, Raleigh, NC.
[‡‡]Physics Department, North Carolina State University, Raleigh, NC.

[1]S. Triambak et al., Phys. Rev. C **95** (2017) 035501.
[2]Cirigliano, Gardner, Holstein, Prog. Part. Nucl. Phys. **71**, 93 (2013).
[3]Gonzalez-Alonso, Naviliat-Cuncic, Severijns, arXiv:1803.08732.

has gathered 300 fb^{-1} at 14 TeV. Shown in red are limits that we could expect from our work, showing that new physics could be discovered in nuclear beta decays.

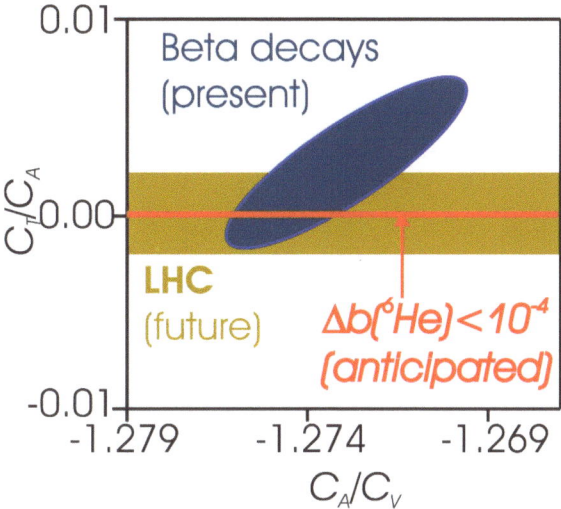

Figure 3.2-1. 90%-C.L. limits on tensor couplings versus the axial-to-vector coupling in neutron decay. Blue tilted ellipse: beta decay data[3], showing the correlation between C_T/C_V and C_A/C_V, generated principally by the neutron-beta-decay data. Brown: upper limits expected by Cirigliano et al. from LHC data assuming no events are detected by the time 300 fb^{-1} are collected at 14 TeV[2]. The red-shaded area is from a spectrum shape determination with uncertainty $\Delta b = 10^{-4}$, the expected sensitivity of the ^{6}HeCRES experiment.

$\beta - \nu$ correlation from laser trapped ^{6}He

Our present effort aims at determining the $\beta - \nu$ correlation a (also called "little-a") by trapping ^{6}He in a laser Magneto-Optical Trap (MOT). The electron is detected via a combination of a multi-wire proportional chamber and a scintillator. The recoiling Li ions are detected via a focusing electric field that guides them onto a position-sensitive Micro Channel Plate (MCP) detector. By measuring the time of flight and the landing position of the Li ion its momentum can be reconstructed. Together the electron and Li momentum can be used for reconstructing the anti-neutrino momentum.

During the past year:

1. The laser-power stabilization system was implemented and MOT shape improved.

2. Monitoring systems for the MOT, including CCD images, taken simultaneously with the little-a data, were implemented.

3. Photo-ion methods for determining the MOT position were carefully studied.

4. Charge distribution measurement showing a remarkable disagreement (19σ) were published[1].

5. Data were taken for a $< 1\%$ determination of the $\beta - \nu$ correlation coefficient.

---
[1]R. Hong et al., Phys. Rev. A **96**, 053411 (2017).

We now can get data for a $\sim 1\%$ determination of the $\beta - \nu$ correlation coefficient in less than 3 days of running time, including calibrations. The experiment runs in a stable fashion. The target sensitivity we have with the present setup is 0.5 %. We expect to finish our systematic analysis and start gathering statistics for the $\beta - \nu$ correlation determination soon.

Shape of the ^{6}He β spectrum using the CRES technique

On a different front, we have established a collaboration to extend the use of the CRES technique, developed by the Project 8 collaboration, to determine the shape of the ^{6}He beta spectrum to determine the strength of the Fierz term b (also called "little-b") and other nuclear physics spectroscopy applications. Design of the system is being finished and we expect to assemble it during the summer of 2018.

3.3 Hardware improvements for the ^{6}He little-a experiment

Y. S. Bagdasarova, K. Bailey[*], C. Cosby, X. Fléchard[†], A. García, B. Graner, R. Hong[*], A. Knecht[‡], T. Larochelle, A. Leredde[*], G. Leum, J. Lu, P. Müller[*], O. Naviliat-Cuncic[§], T. P. O'Connor[*], J. Pedersen, E. Smith, D. W. Storm, and H. E. Swanson

Hardware improvements for the ^{6}He experiment involve calibrations of the high voltage divider monitor system, calibrations of the relative magneto-optical trap (MOT) position with the camera and micro-channel plate (MCP) images, and the construction of new insulation for the liquid nitrogen (LN$_2$) trap for the helium atomic beamline.

The extraction of a from the time-of-flight (TOF) measurement of the ^{6}Li ions relies on accurate modeling of the ion-accelerating electric field in the Monte-Carlo simulation. The electric field is generated by an array of electrodes precsiely held at prescribed high voltages for a desired field configuration. Simulation studies have shown that a is sensitive to the electrode voltage at the level of 8% for a 1% systematic electrode voltage offset. Thus a 0.05% limit in voltage inaccuracy is desired to obtain a 1% measurement in a.

In June 2017, we calibrated the high voltage precision dividers (D1,D2,...D5) used for monitoring the electric field electrode voltages *in situ* using a CPS HVP-250 probe that was cross-calibrated with a NIST probe to 0.02% accuracy (Fig. 3.3-1). The divider readings were sampled at 1.5 minute intervals in their operational ranges and fit against the probe reading using a piecewise linear interpolant $V_{probe}(V_{divider})$. The interpolation error was estimated to be $\Delta V/V < 6 \times 10^{-5}$ and the curves were reproduced ten days later to the level of $\Delta V/V < 6 \times 10^{-5}$. The system remained stable until a sparking event on D5 and D4 in December 2017 caused their readings to change by $\Delta V/V = 9 \times 10^{-3}$ and $\Delta V/V = 6.3 \times 10^{-4}$, respectively. Otherwise, we have verified that from June 2017 to February 2018 the D1-D3

---

[*]Physics Division, Argonne National Laboratory, Argonne, IL.
[†]LPC-Caen, ENSICAEN, Université de Caen, Caen, France.
[‡]Paul Scherrer Institute, Villigen, Switzerland.
[§]Department of Physics and Astronomy and National Superconducting Cyclotron Laboratory, Michigan State University, East Lansing, MI.

calibrations remained accurate to within $\Delta V/V < 1.6 \times 10^{-4}$ (Fig. 3.3-2) and that subsequent calibrations of D4 and D5 showed them to be stable to $\Delta V/V < 5 \times 10^{-5}$. We had previously installed 100k precision resistors between the high voltage output of the supplies and the electrodes in order to filter out high-frequency noise picked up by the MCP. To prevent sparking, the filters for dividers 3-5 were encapsulated in dielectric tubing. In the processes of recalibrating the dividers, the dielectric tubing around the filter resistors was removed, and we observed a significant reduction in voltage spiking in the readout as a result (Fig. 3.3-3). Further, we added 500 pF capacitors to lower the cut-off of the low pass filters to 100 kHz, added copper grounding tape to the dielectric holders to reduce surface charge build up, and replaced D6 with the probe to stop arcing (Fig. 3.3-6). Currently, the electrode voltages are stable and accurate to within 0.02%, meeting our experiment goal.

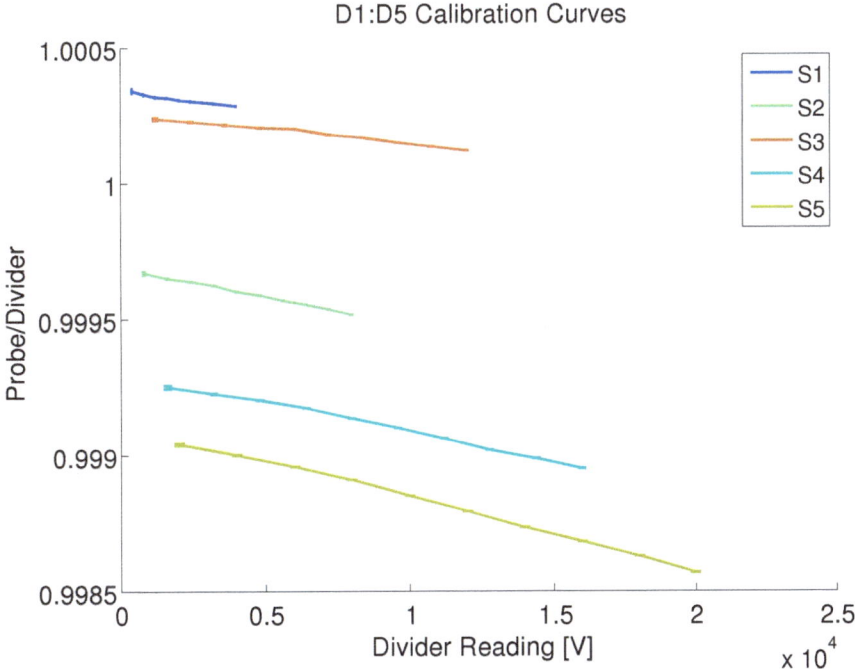

Figure 3.3-1. In situ calibration curves for high voltage dividers used to monitor the electrode voltages, expressed here as the ratio of probe reading (V_{probe}) to divider reading ($V_{divider}$) over the operational ranges of each divider. The linear response of $V_{probe}/V_{divider}$ as a function of voltage is due to the voltage coefficient of resistance for the dividers. It can be read directly from the curves as -0.03 ppm/V, which agrees with the < -0.07 ppm/V spec limit for the dividers.

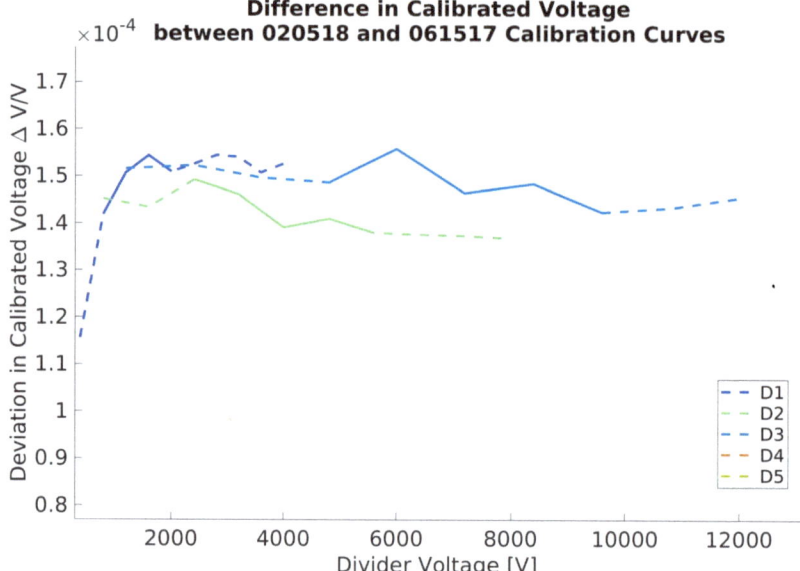

Figure 3.3-2. Change in calibration from June 2017 to February 2018 for dividers 1,2, and 3 expressed in terms of fractional change in probe voltage readout $\Delta V/V$ vs divider voltage readout. $\Delta V/V < 1.6 \times 10^{-4}$

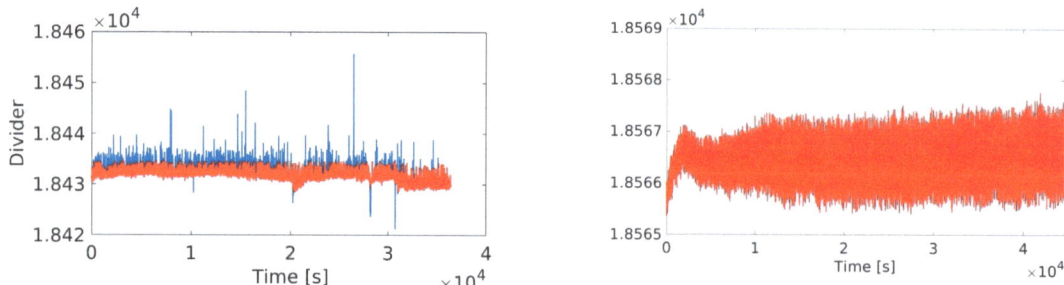

Figure 3.3-3. *Left:* 10-20 V voltage spikes on Divider 5 readout vs time with dielectric tubing in place around filter resistors just before a December 2017 sparking event. The red curve indicates accepted values by a spike filter algorithm. *Right:* Divider 5 readout vs time after dielectric tubing is removed. There is an initial climb of 2 V in the first half hour and $V_{pp} = 1$ V after.

In the last year we developed ^{3}He trapping capabilites and are presently capable of taking ^{3}He, ^{4}He, and ^{6}He photoion data for the purposes of monitoring and measuring the MOT position and the TOF stability. We also developed the ability to simultaneously image the MOT using the CMOS camera while acquiring data. Previously the CMOS camera had required a much brighter MOT than the MCP could withstand. By accumulating exposures of the dimmer MOT over several minutes we are able to correlate the MCP image and TOF data with the camera images in time (see (Sec. 3.4) - Fig. 3.4-1).

New ruler face prototypes for the CMOS camera image calibration were constructed using CENPA's circuit board laser etcher. The CMOS camera calibration involves imaging a precisely etched 500 μm ruler grid in the focal plane of the camera in order to obtain a pixel to mm calibration function[1]. The new ruler grids are etched into sheets of anodiized aluminum. These faces are meant to be glued over the original 20 × 35 mm face region of the ruler structure. Analysis of the ruler image shows the 250 μm grid to be easily identifiable by software and to be uniformly spaced to 6 μm (Fig. 3.3-4). A 500 μm ruler was also built for a future camera calibration.

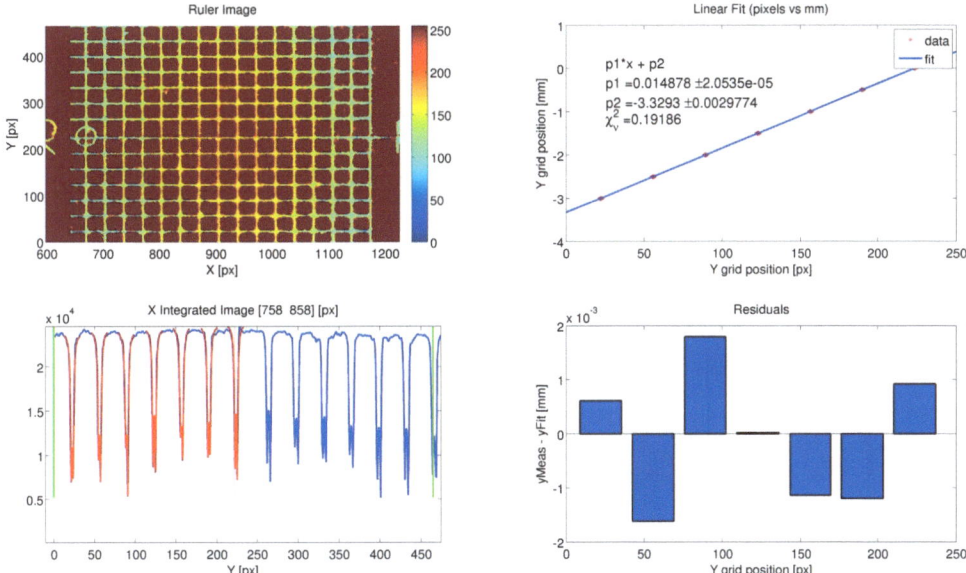

Figure 3.3-4. Analysis of new ruler face grid. *Top-left:* Microscope image of the ruler grid in X and Y pixels. *Bottom-left:* Projections of grid lines along the Y axis are fit to Lorentzians to obtian grid mark positions in pixels. *Top-right:* Linear fit of the imaged grid parks in pixels to their ideal positions in mm. *Bottom-right:* The raw residuals of the linear fit reveal uniformity to better than 6 μm.

The cannister insulation for the LN$_2$ cooling of the discharge source was redesigned and constructed (Fig. 3.3-5). The RF-discharge tube is the first stage of trapping ^{6}He, where ^{6}He atoms are excited to the metastable state by way of collisions with electrons in the RF-generated Xenon plasma. The tube is cooled in order to produce a slower and tighter velocity distribution of ^{6}He metastables for more efficient downstream trapping. The new cannister insulation has solved the problem of LN$_2$ leaking into the discharge source and has increased the run time of a single dewar from 8 hours to up to 22 hours without interruption.

---

[1]CENPA Annual Report, University of Washington (2017) p. 53.

Figure 3.3-5. The improved insulation for the liquid nitrogen cooling for the discharge source. *Left:* Exposed cannister with inner insulation in place. *Right:* Insulation sealed off with cryogenic vapor barrier tape.

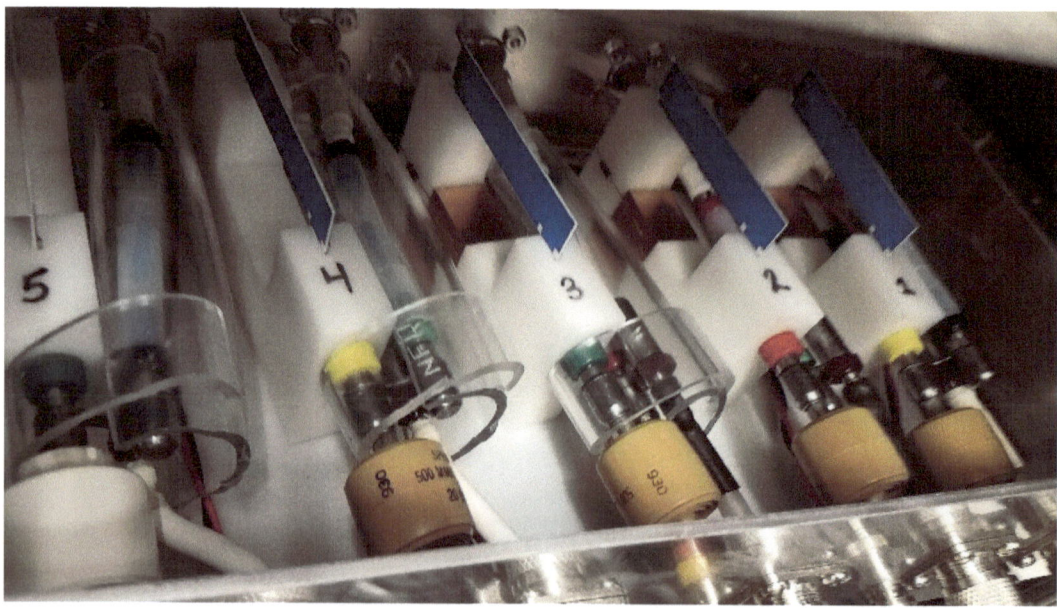

Figure 3.3-6. The high voltage precision dividers for electrode voltage monitoring system and the low-pass filters. Filters for dividers 3-5 are initially encapsulated in dielectric tubing to prevent sparking.

In June 2017 we acquired ^{6}He decay data concurrently with ^{6}He photoion data at full and half electric field. We ran for a total of 54 hours at an average triple coincidence rate of 4 Hz, acquiring statistics for a 1.6% measurement of a in both data sets. We also took TOF measurements of the ^{3}He, ^{4}He, and ^{6}He photoions while scaling the electric field for the purpose of diagnosing systematics.

3.4 Progress in the analysis of the ^{6}He little-a experiment

Y. S. Bagdasarova, K. Bailey*, C. Cosby, X. Fléchard[†], A. García, B. Graner, R. Hong*, A. Knecht[‡], T. Larochelle, A. Leredde*, G. Leum, J. Lu, P. Müller*, O. Naviliat-Cuncic[§], T. P. O'Connor*, J. Pedersen, E. Smith, D. W. Storm, and H. E. Swanson

In June 2017 we took additional data under stable trap and electric field conditions in two different field geometries for a measurement of a to the 1.5% level. The initial analysis of these data, based on previous understanding of the detector systems, has been performed by fitting the time-of-flight (TOF) spectrum with an Monte-Carlo simulation of the experiment where a is a fit parameter. The resulting fit of the TOF spectra are poor, and there are clear deviations between simulation and data that cannot be explained by our current modeling of the system. This year has therefore been devoted to experimentally verifying our parameters and systematic uncertainties. Major efforts have been applied to verifying the stability of the magneto-optical trap (MOT) position and micro-channel plate (MCP) timing response using CMOS camera imaging, MCP imaging of the Penning ions, and TOF measurements of the ^{4}He and ^{3}He photoions. The MOT position was repeatedly confirmed to be stable to 10-15 μm over 4 hours on both the CMOS camera image and the MCP image (Fig. 3.4-1), while the photoion TOF was confirmed to be stable to 50 ps over 20 hours (Fig. 3.4-3). These numbers satisfy the stability requirements of the experiment.

The absolute position of the MOT is one of the largest systematics of the experiment, leading to a systematic shift of 1.5%/100 μm in a. The absolute position and timing response of the detectors are strongly correlated in a, so independent calibrations of one or both is necessary to obtain a 1% measurement.

Effort has been put into characterizing the timing response of the MCP for ^{6}He decay events and photoion events in terms of other measurable parameters. The time of flight of an event is defined as $TOF \equiv T_{MCP} - T_{PMT} - T_0$, where T_{MCP} and T_{PMT} are the readout times for events that trigger the MCP and the photomultiplier tube (PMT), respectively, and T_0 is the correction for delays in the detector system. We previously observed T_0 to correlate significantly with the MCP position and charge amplitude, particularly in the timing peak from backscattered β-rays and decays originating from the MCP, varying by 500 ps over

---

*Physics Division, Argonne National Laboratory, Argonne, IL.
[†]LPC-Caen, ENSICAEN, Université de Caen, Caen, France.
[‡]Paul Scherrer Institute, Villigen, Switzerland.
[§]Department of Physics and Astronomy and National Superconducting Cyclotron Laboratory, Michigan State University, East Lansing, MI.

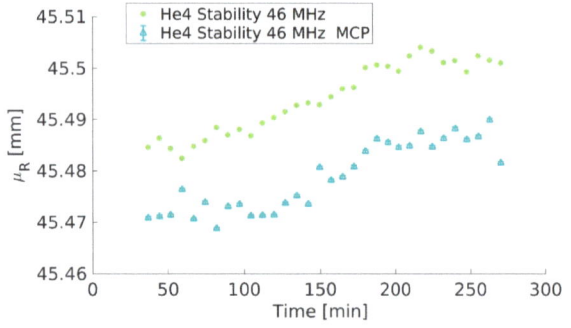
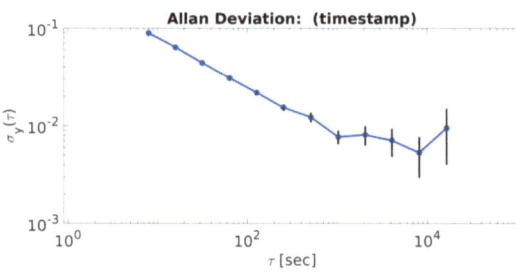

Figure 3.4-1. ^{4}He MOT transverse position vs time determined with simultaneous CMOS camera imaging (green) and MCP imaging (turquoise), showing 15 μm stability over 4 hours.

Figure 3.4-2. Allan deviation of ^{4}He photoion TOF shows long-term stability.

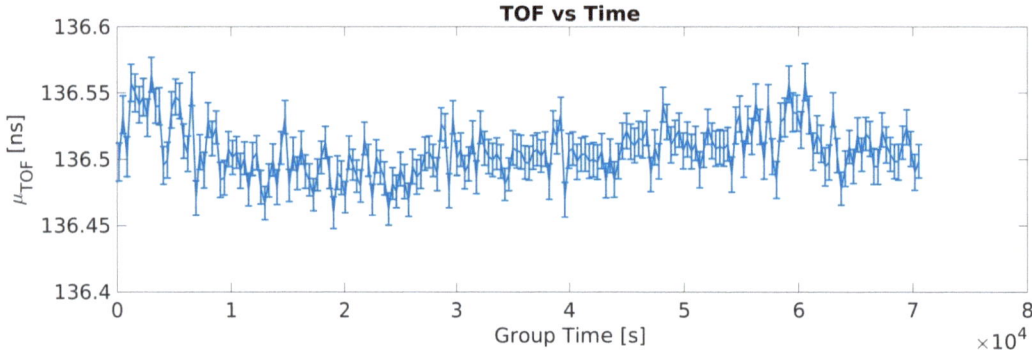

Figure 3.4-3. ^{4}He photoion TOF vs time over 20 hours, showing stability to within 50 ps.

80 mm of the MCP[1]. This year we used this analysis to develop a second-order polynomial correction for the ^{6}He decay data. We found that our correction has little effect on the fit of a. Additionally, it cannot explain other strange timing effects observed for the photoion data, localized within a 1 mm region at the center of the MCP. This exhibits a 800 ps difference in the reconstruction of T_0 from ^{6}He, ^{4}He and ^{3}He photoion TOFs when we vary the electric field strength from 100% to 35%. The 800 ps variation in T_0 as a function of field for the photoions does not meet the calibration requirement of 20 ps for the determination of the absolute MOT position using this method. To test the effect of particle type on MCP timing response we are currently investigating alternate methods for calibrating the MCP timing.

The alignment and beam profile of the ionizing laser for the photoions was reexamined multiple times this year. For the June 2017 data run we achieved a 3 mm flat-top profile that resulted in photoion TOF variations below 25 ps in the region of overlap with the MOT. In October 2017, the profile and alignment were reexamined. Physical inspection of the beam confirmed the desired size and shape of the profile while the measured TOF trends as a

---

[1]CENPA Annual Report, University of Washington (2017) p. 57.

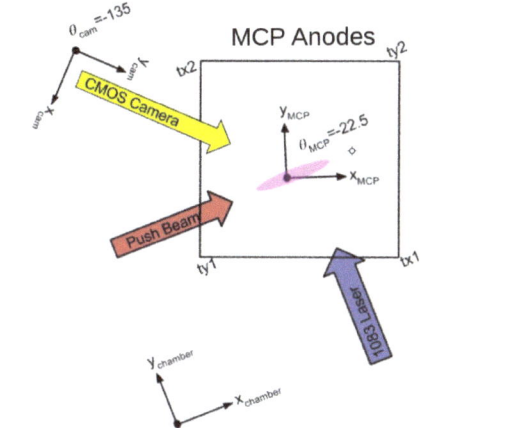

Figure 3.4-4. MCP, camera, and chamber transverse coordinate system orientations.

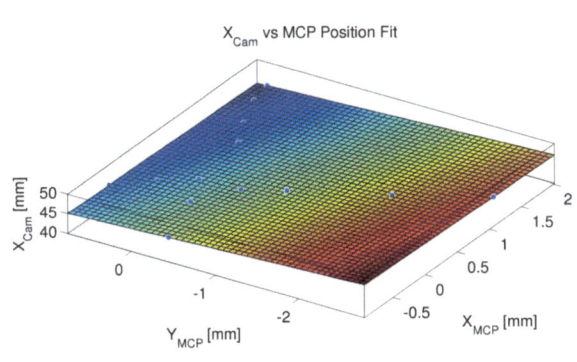

Figure 3.4-5. Fit of transformation function $X_{cam}(X_{MCP}, Y_{MCP})$, where X_{MCP} and Y_{MCP} are centroids of the Penning ion Gaussian image seen by the MCP and X_{cam} are the corresponding centroids of the camera images.

function of laser position on the MOT confirmed a constant TOF within 50 ps in the region but did not confirm a flat-top shape.

A combination of the camera and MCP image allows us to image all three dimensions of the MOT. The camera images the MOT directly along the vertical and transverse directions, while the MCP images the MOT indirectly along the plane of the chamber via Penning ion events. The transverse axis of the camera is rotated with respect to the MCP axes as shown in Fig. 3.4-4. To check the agreement of the two images for the transverse position, the MOT was moved in the horizontal directions by several millimeters. A fit of the transformation was obtained to 100 μm accuracy (Fig. 3.4-5). Both MCP and camera images also revealed reproducible jumps on the order of hundreds of μm in the motion of the MOT on top of the expected smooth motion from varying the magnetic field (Fig. 3.4-6). These jumps occurred both along the primary direction of motion and the orthogonal directions. The limited accuracy of the current supplies would only account for 17μm of the error. A possible explanation for the distortions is that the trapping laser profiles form interference patterns around the vicinity of the MOT, but this remains unconfirmed. This does not effect the determination of a since the MOT will remain stationary during the measurements.

To diagnose the photoion timing issues, in November and December of 2017, we experimentally measured the correlation between the MOT vertical position Z and the ^{4}He and ^{3}He photoion TOFs. To do this we changed the MOT position by several millimeters by varying the magnetic field for the trap while imaging with the camera, and subsequently measured the photoion TOF as a function of Z. Table 3.4-1 lists the measured slopes $dTOF/dZ$ and reduced χ^2 values for 3 independent trials. The agreement with the expected values from simulation are within 100 ps/mm. As Fig. 3.4-7 shows, large residuals lead to poor reduced

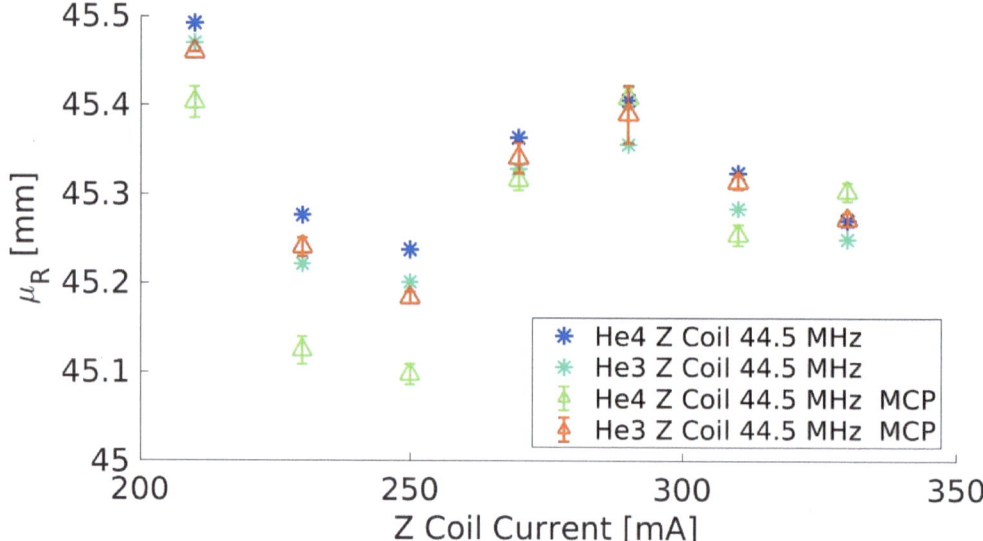

Figure 3.4-6. Hundred μm kinks in the the transverse position seen by MCP and CMOS camera as a function of moving the MOT in the vertical direction.

χ^2 values for many of the fits. Limited accuracy of the current control for the magnetic fields could only account for 12 ps of the error. This indicated an inconsistency in the measured TOF of the photoions that again may be due to distortions in MOT profile as a function of trap position, variations in the MCP timing response as a function of position or magnetic field, or a combination of all factors.

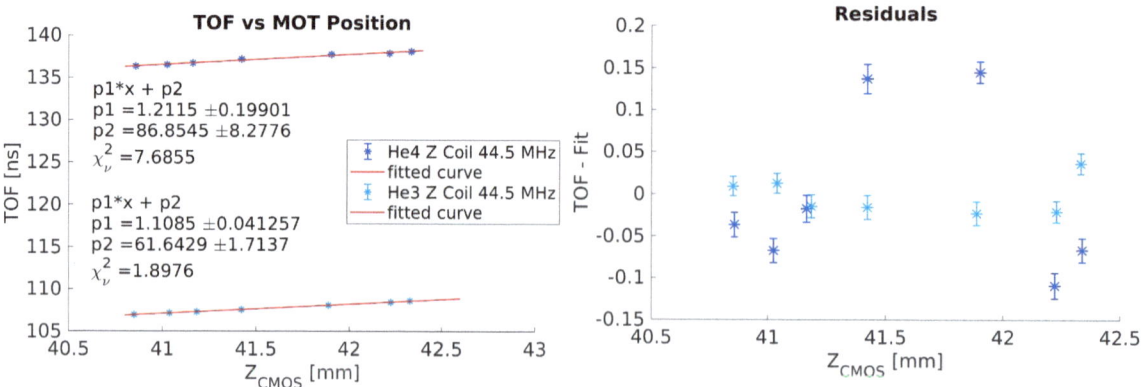

Figure 3.4-7. *Left:* ^{4}He and ^{3}He photoion TOF vs MOT Z position fit for 12/14/2017. *Right:* Fit residuals show unexplained 150 ps fluctuations.

Isotope		12/14/2017	12/12/2017	11/27/2017	Simulation
^{4}He	$d\mathrm{TOF}/dZ$ [mm/ns]	1.21 ± 0.20	1.20 ± 0.04	1.34 ± 0.03	1.21 ± 0.01
	χ^2_ν	7.7	1.5	0.5	
^{3}He	$d\mathrm{TOF}/dZ$ [mm/ns]	1.11 ± 0.04	1.08 ± 0.08	0.97 ± 0.11	1.03 ± 0.01
	χ^2_ν	1.9	3.8	3.4	

Table 3.4-1. Fits of $d\mathrm{TOF}/dZ$ for ^{4}He and ^{3}He in three independent trials.

The MCP timing uncertainties mentioned above make the determination of the MOT distance to the MCP via the photoion TOF measurements extremely challenging for our configuration. We are currently exploring a more direct method for calibrating the MCP timing response with ^{249}Cf, a source that emits coincident α- and γ-rays. For our final checks, we will use the simulation to reassess our model parameters and systematic uncertainties of our detection scheme for the high and low field data taken in June 2017.

3.5 Progress towards precision measurement of the ^{6}He β-decay spectrum via cyclotron radiation emission spectroscopy

K. Bailey*, R. Buch, W. A. Byron, M. Fertl, R. Farrell, A. García, G. T. Garvey, B. Graner, M. Guigue, K. S. Khaw, N. S. Oblath[†], P. Mueller*, J. Pedersen, R. G. H. Robertson, G. Rybka, G. Savard*, E. Smith, D. Stancil[‡], H. E. Swanson, J. Tedeschi[†], B. A. VanDevender[†], F. Walsh, F. Wietfeldt[§], and A. R. Young[¶]

Precise measurement of the energy spectrum of ^{6}He β-decay provides a method of probing the nucleus for chirality-flipping interactions that fall outside the scope of Standard Model (SM) physics. In 2017 our group began a ^{6}He β-decay spectrum measurement using the technique of cyclotron emission radiation spectroscopy (CRES). In a CRES experiment, the kinetic energy of each decay electron can be measured to high precision by detecting the frequency of cyclotron radiation emitted as the particle orbits in a known magnetic field. The CRES technique was first developed at CENPA by the Project 8 collaboration and has recently been successfully demonstrated in the decay of krypton atoms at relatively low energy[1]. Recent progress largely consists of the design work for the energy spectrum measurement.

Decay volume

At the center of the experiment is a decay volume, pictured in Figure Fig. 3.5-1, formed by a cylindrical waveguide oriented along the primary magnetic field axis, with one or more

---

*Argonne National Laboratory, Lemont, IL.
[†]Pacific Northwest National Laboratory, Richland, WA.
[‡]Department of Electrical Engineering, North Carolina State University, Raleigh, NC.
[§]Department of Physics, Tulane University, New Orleans, LA.
[¶]Department of Physics, North Carolina State University, Raleigh, NC.

[1]D. M. Asner *et. al.*, Phys. Rev. Lett. **114**, 162501 (2015).

magnetic trap coils wound around the waveguide's outer surface. ^{6}He atoms produced in the CENPA tandem linear accelerator will be piped into the decay volume, where the energy spectrum measurement takes place. Electrons emitted almost perpendicular to the combined magnetic field/waveguide axis (designated the $\hat{z}$ axis) will undergo cyclotron orbits and radiate into one or more guide modes. Because the electrons have a conserved orbital magnetic dipole moment, they will be confined to a local field minimum. One or more trapping coils create an additional axial field designed to trap electrons emitted at pitch angles close to $\pi/2$ relative to the $\hat{z}$ axis. Electrons emitted with substantial momentum along the $\hat{z}$ axis will escape the trap and be lost on the waveguide walls or the kapton windows separating the decay volume from the rest of the waveguide.

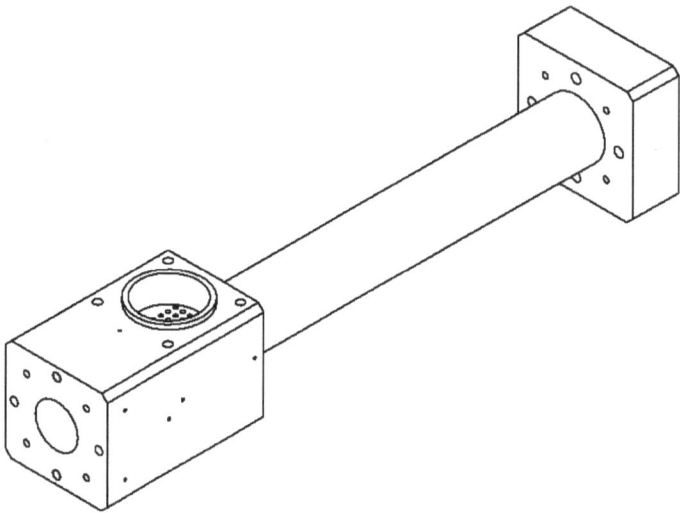

Figure 3.5-1. A 3D rendering of the decay volume for the ^{6}He β-decay energy spectrum measurement. Helium atoms will be compressed by a turbo-molecular pump and enter the hollow cylindrical guide through the array of small holes on the top surface partially visible on the lower left. These holes are much smaller than the wavelength of the cyclotron radiation and therefore do not degrade the guide performance. Additional magnet coils will be wound on the outside of the guide to trap electrons emitted with sufficiently small momentum along the $\hat{z}$ axis, which will then oscillate back and forth along the guide axis as they continue to emit cyclotron radiation. The emitted radiation will feature a characteristic increase in frequency as the electron loses energy; the microwave radiation will be transmitted to amplifiers on either end of the cylindrical guide by linear microwave polarizers and standard rectangular WR42 waveguides.

Magnet system

The decay cell is designed to be inserted into the bore of a 7 Tesla superconducting American Magnetics model 8946 system furnished by collaborators at Argonne National Laboratory. The magnet has a 5" wide, 40" long central bore oriented horizontally. According to the manufacturer's specifications, the central field spatial non-uniformity is below the level of 1 ppm per millimeter, which is more than adequate for the present purpose. The magnet, pictured in Fig. 3.5-2, was shipped to CENPA in February 2018.

Figure 3.5-2. J. Pedersen stands next to the magnet system to be used in the ^{6}He β-decay energy spectrum CRES measurement.

Waveguide design

The signal power to be obtained from individual electrons is on the order of one femtowatt (10^{-15} W), which is much lower than the thermal noise power of room-temperature radio frequency (RF) amplifiers. The RF amplifiers for the ^{6}He CRES measurement must therefore be operated near 4 Kelvin to achieve an acceptable signal-to-noise ratio (SNR). However, the temperature of the guide walls does not contribute to the thermal noise in the RF system insofar as they are perfect conductors. We thus decided to change the design temperature of the decay cell to enable testing the system with krypton, which freezes at 120 K. Our Project 8 collaborators at CENPA advised us not to place an RF short on the side of the decay cell farthest from the RF amplifiers. Replacing the initial Project 8 reflective RF waveguide short with a high-temperature terminator would have contributed excessive amounts of blackbody radiation, so the design was altered to form a 'U' shape with the active or lossy RF components at either end (nearest the cryocooler) and the decay volume near the center (see Fig. 3.5-3). The compactness of this design carries the added advantage that the high-field region does not require extensive temperature control, and the magnet can be slid back and forth horizontally on its axis parallel to the straight sections of the waveguide 'U'. This enables us to assemble the entire vacuum system before putting the magnet in place, which gives a substantial advantage in ease-of-operation over the initial hardware design presented previously[1].

---

[1]CENPA Annual Report, University of Washington (2017) p. 62.

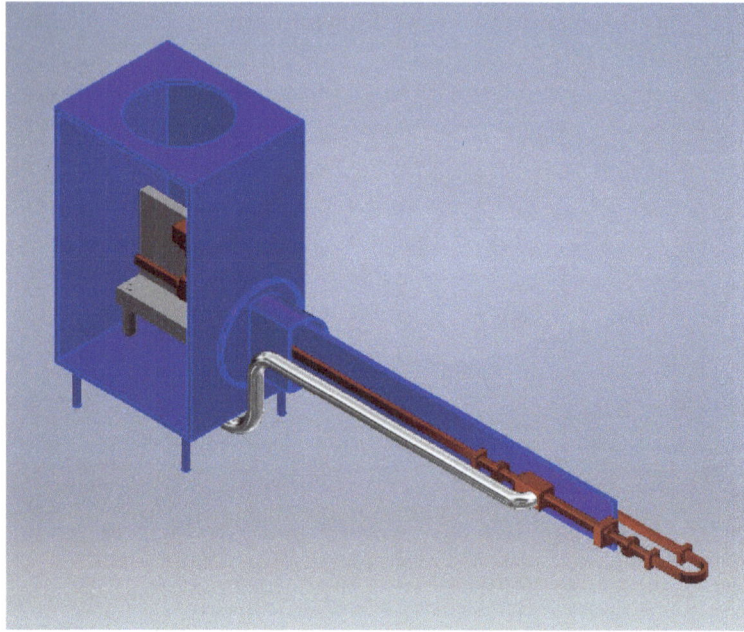

Figure 3.5-3. A partial rendering of the RF waveguide components and surrounding parts. Components at 4 K reside inside the 40K copper shield, indicated in blue, while the waveguide components are shaped into a 'U' that straddles a projecting low temperature bus bar. One terminator and one amplifier can be seen partially enclosed in the 40 K box, which is open at the top to allow insertion of the second stage cryocooler. Photonic crystal flanges are used at points where the WR42 waveguides pass into the 40 K box to minimize heat transfer. The titanium pipe conducts ^{6}He into the decay volume, with trap coils wound on axis.

Cryogenic system

The design is broken into two basic temperature regimes following the division between the first (40 K) and second stage (4 K) of the cryocooler: The low-temperature components are enclosed inside a copper shielding box wrapped with multilayer insulation and fixed to the temperature of the first cryocooler stage. This prevents the parts at or near 4 K from absorbing an excessive amount of black body radiation from room temperature surfaces. The high-temperature (above 40 K) components are fit into a 4-inch diameter space for insertion into the high-field region.

Simulation

Simulations of the electron trajectories are carried out using the Kassiopeia package developed by the KATRIN collaboration, while the waveguide response and received power are simulated using the Katydid package developed by Project 8. This enables us to optimize the trapping coil design and the pitch angle acceptance window to determine systematic sources of error, such as the Doppler effect on the cyclotron radiation from a moving electron, power leakage into non-propagating modes of the waveguides, and electron collisions with the guide walls.

Data acquisition

Data will be acquired and digitized using a ROACH2 system developed by the CASPER collaboration for use in high-bandwidth, low SNR applications in radio astronomy[1]. The ROACH2 is based on a Xilinx Virtex6 field-programmable gate array (FPGA), which is optimized for high rate digital signal processing applications (see Fig. 3.5-4). The FPGA and associated software licenses were generously furnished to CENPA at no cost through the Xilinx University Donation Program. The FPGA takes its input from a 5 Gigasample/second analog-to-digital converter (ADC), and can be configured to output IPV4-packets of data through 4 parallel 10 GbE interfaces.

Figure 3.5-4. A photo of the inner parts of the ROACH2 FPGA system to be used to digitize cyclotron radiation signals with a bandwidth of 2 GHz. The 5 Gs/sec ADC is visible in the foreground, with the FPGA board and quad 10 GbE card at the back. Analog signals enter via the SMA bulkheads on the backplane, seen on the right.

Storage and analysis of the data output from the ROACH system is perhaps the largest remaining area of the experiment yet to be considered in depth. The ADC used can generate up to $5 \cdot 10^9$ 8-bit samples per second. If raw time-domain data were to be stored with some lossless format, this would require 5 gigabytes of storage per second of live time. We intend to address this issue by filtering and storing frequency domain data with some combination of limited bandwidth, finite resolution, downsampling, and acceptable dead time. We also intend to make use of the local CENPA computing cluster, recently upgraded with hardware acquired from the National Energy Research Scientific Computing Center. The upgraded cluster is expected to have 760 terabytes of available storage.

---

[1] http://www.casper.berkeley.edu

4 Precision muon physics

MuSun

4.1 Muon capture and the MuSun experiment

D. W. Hertzog, P. Kammel, E. T. Muldoon, D. J. Prindle, R. A. Ryan, and D. J. Salvat

Muon capture provides a powerful tool to study properties and structure of the nucleon and few nucleon systems as predicted by effective theories (EFT) founded on Quantum Chromodynamics. Our program focusses on capture from the simplest of all muonic atoms, namely the theoretically-pristine muonic hydrogen (MuCap experiment) as well as muonic deuterium (MuSun experiment). Our collaboration has pioneered a novel active-target method based upon the development of high-pressure time-projection chambers (TPC) filled with hydrogen/deuterium gas, which reduced earlier experimental uncertainties by about an order of magnitude.

Our MuCap experiment[1] has measured the singlet muonic hydrogen capture rate, $\Lambda_{\text{singlet}}^{\text{MuCap}} = 715.6(7.4)\ s^{-1}$, with unprecedented precision, to determine the induced pseudoscalar nucleon coupling g_P. In the past, the uncertainty in Λ_{singlet} introduced by the momentum dependence of the other weak form-factors was considered negligible. However, a recent determination of the axial-vector charge-radius squared from a new model-independent z-expansion analysis[2] of neutrino-nucleon scattering data determined $r_A^2(z\ \text{exp.}) = 0.46(22)\ \text{fm}^2$, with a more realistic, but 10-fold increased, uncertainty compared to previous model-dependent analyses. Given this dramatically-increased uncertainty of the axial-vector form factor $F_A(q^2) = F_A(0)\left(1 + \frac{1}{6}r_A^2 q^2 + \ldots\right)$, we studied the following questions in a recent review[3], and answered both in the affirmative. Does the comparison of g_P between experiment and theory still provide a robust test of EFT and QCD? And, in a reversal of strategy, can muon capture be used to determine a competitive value of r_A^2 ? The review updates g_P^{MuCap} to 8.23(83), which implies $g_P^{\text{theory}}/g_P^{\text{MuCap}} = 1.00(8)$, confirming theory at the 8% level. If instead, the predicted expression for g_P^{theory} is employed as input, then the capture rate alone determines $r_A^2(\mu H) = 0.46(24)\ \text{fm}^2$, or together with the independent recent analysis of νd data, a weighted average $r_A^2(\text{ave.}) = 0.453(23)\ \text{fm}^2$. The importance of an improved r_A^2 determination for phenomenology is illustrated by considering the impact on critical neutrino-nucleus cross sections at neutrino oscillation experiments.

One of the most interesting topics for muon capture in the few-body sector is the family of two-nucleon weak-interactions processes. In these reactions only a single unknown low energy constant (LEC) enters the description up to the required order, which characterizes

---

[1]V.A. Andreev *et al.*, Phys. Rev. Lett. **110**, 012504 (2013).
[2]A S. Meyer *et al.*, Phys. Rev. D **93**, 113015 (2016).
[3]R. Hill, P. Kammel, W. Marciano and A. Sirlin, 2018, Rep. Prog. Phys. in press https://doi.org/10.1088/1361-6633/aac190

the strength of the axial-vector coupling to a four-nucleon vertex, the two-nucleon analog to g_A for the nucleon. This family contains muon capture on the deuteron,

$$\mu + d \rightarrow n + n + \nu, \qquad (1)$$

together with astrophysics reactions of fundamental interest, in particular, pp fusion, which is the primary energy source in the sun and the main sequence stars, and the νd reactions, which provided the evidence for solar neutrino oscillations at the Sudbury Neutrino Observatory (SNO). The extremely small rates of these processes do not allow their quantitative measurement under terrestrial conditions; they can only be calculated by theory, with information derived from the more-complex three-nucleon system. MuSun plans to determine the rate Λ_d of reaction (1) to 1.5%, where Λ_d denotes the capture rate from the doublet hyperfine state of the muonic deuterium $1S$ state. Current experiments are at the 6-10% level and the most-precise one disagrees with the latest 1% theory calculation by more than 3-sigma. The LEC will be determined at the 20% level, i.e. 5 times better than what is presently known from the two-nucleon system.

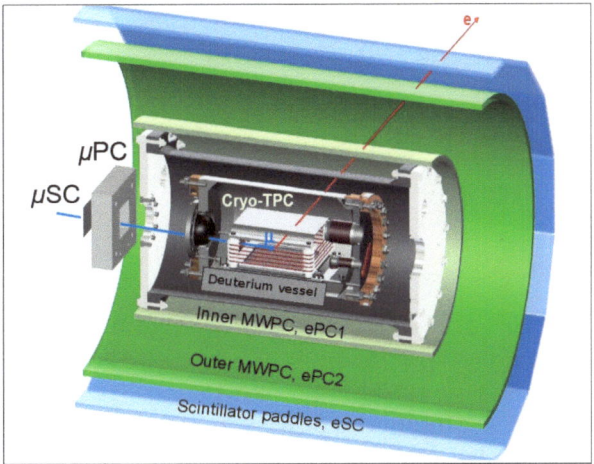

Figure 4.1-1. MuSun detector model. Muons pass through entrance detectors to stop in a deuterium target TPC. Tracking in the TPC selects muons stopping in the fiducial volume with sufficient distance to chamber walls. Decay electron are detected in two cylindrical wire chambers (green) and a 16-fold segmented scintillator array. The lifetime is determined from the measured time difference between the fast muon entrance detector and the decay electron scintillator array.

The MuSun experiment, shown in Fig. 4.1-1, uses the so-called "lifetime method". The disappearance rate, $\lambda_{\mu d}$, of negative muons in an active deuterium target is measured. The capture rate is then determined as the difference $\Lambda_d \approx \lambda_{\mu d} - \lambda_{\mu^+}$, where λ_{μ^+} is the precisely-known positive muon decay rate[1]. A key aspect for achieving the required precision was the development of a new cryogenic high-density TPC operating with ultra-pure deuterium. Tracking in three dimensions in the TPC eliminates the experimental impact of most muon-stops in wall material. Gas purity at the 10^{-9} level suppresses transfer to impurity elements,

---

[1]V. Tishchenko et al., Phys. Rev. D **87**, 052003 (2013).

where capture occurs with a much higher rate than in deuterium. The target conditions ($T = 31\ K$ and density 6.5% of liquid-hydrogen density) are optimized to allow an unambiguous extraction of Λ_d, independent of muonic atomic-physics complications that occur after muons stop in deuterium, in particular the numerous dd-fusion reaction catalyzed by the muons[1]. The high gas density implies that the cryo-TPC does not have internal gas gain and that drift voltages of 80 kV are required.

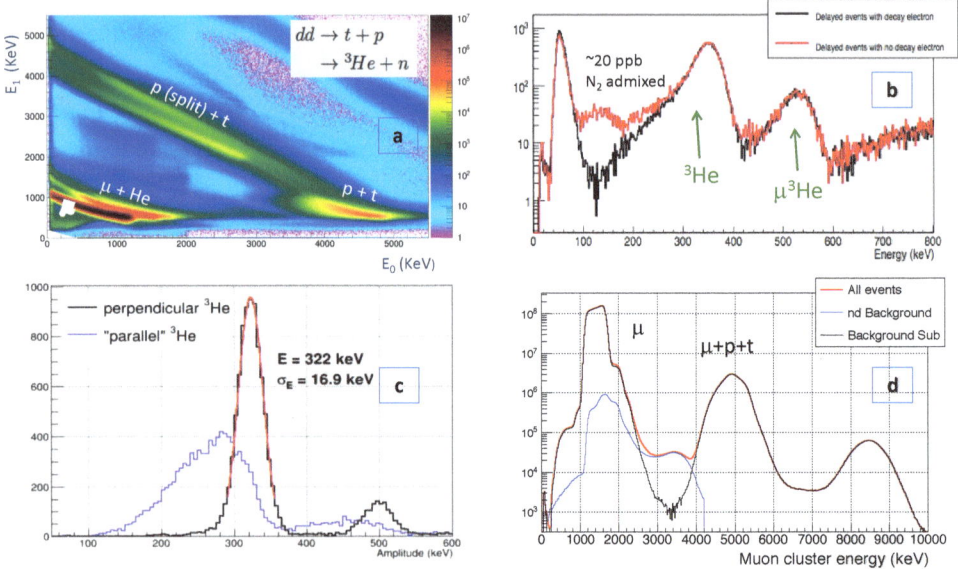

Figure 4.1-2. Analysis examples: (a) Energy deposition in the muon stop-pad (E_0) vs. penultimate pad (E_1) discerns pure muon stops with those overlapping with fusion recoils, (b) nuclear recoils from nitrogen capture can be separated from $^3He + n$ fusions, (c) the low energy tail from ^3He is explained by recombination losses for recoils parallel to the drift field. (d) The yield of pt-fusions, required for the fusion correction, is determined at the 10^{-3} level with the background shape constructed from data.

After final hardware upgrades MuSun collected the full statistics of 1.4×10^{10} events in two main production runs R2014 and R2015, followed, in 2016, by a shorter systematics run. MuSun is a high statistics/high precision experiment, which needs supercomputer resources to analyze the several hundred TB of primary and processed data (provided by an XSEDE[2] grant). The UW team led the development of the staged-analysis framework and maintains it. We upgraded the MuSun Monte Carlo to a well-tuned and indispensable analysis tool, and run large production sets. The analysis of the high quality R2014 data is well under way (Ph.D. Thesis R. Ryan). A data driven method to quantify a key systematic effect, the distortion of measured time spectra by muon catalyzed fusion reactions, was recently

---

[1]P. Kammel and K. Kubodera, *Ann. Rev. Nucl. Part. Sci.*, 60:327–353, 2010.

[2]The Extreme Science and Engineering Discovery Environment is supported by the National Science Foundation.

developed. The R2016 data (Ph.D. Thesis E. Muldoon) was successfully processed through the whole analysis chain, and shows stable and consistent results over the course of the run. The much-improved TPC resolution and gas chromatography allows us to monitor gas purity in-situ at the 10^{-9} level. Several analysis examples are shown in Fig. 4.1-2.

4.2 New data-driven method to quantify fusion-induced distortions in the MuSun experiment

D. W. Hertzog, P. Kammel, E. T. Muldoon, D. J. Prindle, R. A. Ryan, and D. J. Salvat

Analysis of the R2014 MuSun dataset, comprising half the full statistics required to reach our precision goal, is in an advanced state. The 6×10^9 events have been subdivided into smaller datasets, determined by running conditions, in order to ensure stability in the observed lifetime across the entire beam period. A high-statistics lifetime fit with normalized residuals can been seen in Fig. 4.2-1. The residuals show good agreement with the exponential fit, except for times before 1 μs, where muon-catalyzed fusion interference distorts the lifetime. A new method has been developed to correct for this challenging systematic.

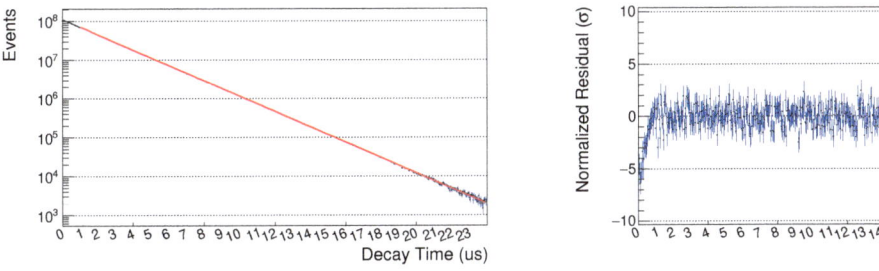

Figure 4.2-1. An exponential fit of the full R2014 dataset (left) and the corresponding fit residuals (right). The fit residuals show good agreement between the fit function and the data, except for times before $1\mu s$, where muon-catalyzed fusion effects lead to a distortion of the observed lifetime.

Resonant formation of $dd\mu$ molecules within the target gas of the MuSun experiment quickly result in muon-catalyzed fusion processes, producing either a neutron and ^{3}He or a proton and a triton. The muon is recycled after the fusion, and can undergo normal decay. Although these processes have no effect on the muon lifetime measurement directly, the presence of fusion products in the TPC can lead to a mis-reconstruction of the muon-stop position, particularly for the proton which has a long range in the target gas. If a muon-stop inside the fiducial volume is mis-reconstructed outside of the fiducial volume, the event is cut due to the presence of the fusion products. Due to the non-exponential time distribution of events proceeded by a fusion, these 'migration' events lead to a distortion of the measured lifetime.

We expect a fraction, ϵ, of events which catalyze a fusion. Due to track mis-reconstruction, migration events will be artificially accepted or rejected by the fiducial-volume cut, resulting in an observed fusion fraction, $\tilde{\epsilon}$. The shift in the measured lifetime is given by

$$\Delta\lambda = \kappa(\epsilon - \tilde{\epsilon}) \qquad (1)$$

In order to obtain the true expected fusion fraction, ϵ, a version of the distribution in the absence of fusion migrations must be obtained. In principle, migrations occur only on the fiducial volume boundary, and primarily in the Z direction. As such, one can suppress the migrations by depleting the stops around the fiducial volume boundary. No cuts can be made on the stop position, as this is already coupled to fusion interference products. However, by making cuts on the entrance location of the muon, the stop distribution can be shaped to reduce events on the fiducial volume boundaries, as shown in Fig. 4.2-2.

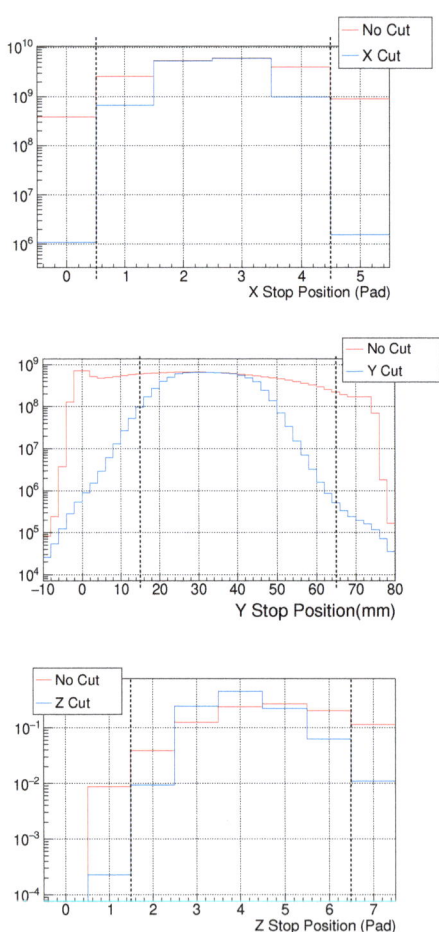

Figure 4.2-2. Stop distributions in (*top*) X, (*middle*) Y, and (*bottom*) Z with and without cuts made on the first pulse in the TPC track. For each distribution, the fiducial volume cut is denoted with dashed lines. The cuts reduce the number of stops on the fiducial volume boundary in each direction, demonstrating the ability to suppress migration events and obtain a dataset free from muon-catalyzed fusion effects.

The constant of proportionality, k, is obtained by scanning the z slices of the TPC, which have various amounts of fusion migration due to the difference in the distribution of stops in each row.

The results of the correction for each of the five sub-datasets, as well as the sum for the R2014 data can been seen on the left in Fig. 4.2-3. A lifetime shift of ~ 23 s^{-1} is found for all datasets. The new approach was also verified with Monte Carlo generated data, see Fig. 4.2-3. The first bin shows the injected/true value of the Monte Carlo. The second and third bins show the fitted lifetime before and after correction, respectively. A lifetime shift of 10 s^{-1} is found for the Monte Carlo correction, resulting in a value within 3 s^{-1} of the truth value.

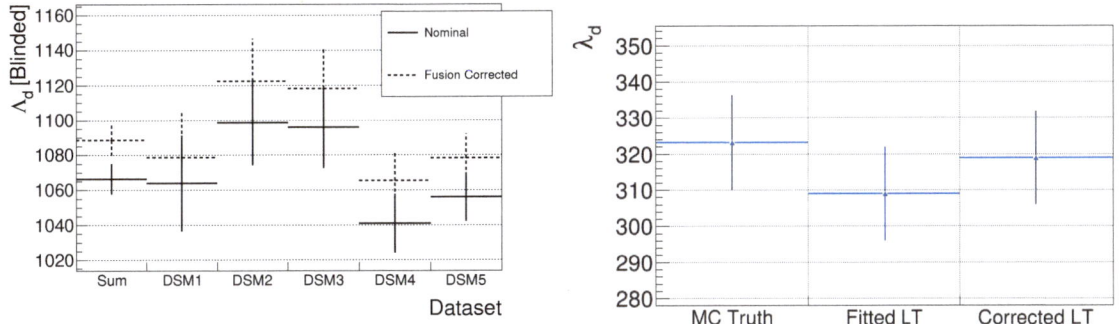

Figure 4.2-3. Fusion correction to the lifetime for (*left*) the sum and sub datasets of the R2014 data and (*right*) the MuSun Monte Carlo simulation. Although the approach is data driven, it is verified using Monte Carlo. When applied to production data, a lifetime shift of 23 Hz is obtained for the R2014 dataset.

4.3 MuSun 2015 data analysis

D. W. Hertzog, P. Kammel, E. T. Muldoon, D. J. Prindle, R. A. Ryan, and D. J. Salvat

The MuSun 2015 production run collected 7 weeks of μ^- and 1 week of μ^+ data, resulting in 50 TB of data which are stored at the Texas Advanced Computing Center (TACC) Ranch mass-storage facility.

In 2017 a preliminary analysis was completed over the highest quality data from the 2015 production run. This was our first analysis pass using the newly-upgraded TACC Stampede 2 computing cluster. In this iteration, we incorporated substantial upgrades to the MuSun analysis software. The MuSun analysis previously involved two stages: Stage one processes raw detector output to reconstruct physical objects, such as particle tracks. Stage two performs more physically-motivated analyses and produces muon decay-time histograms. Stage one was initially incompatible with the new system; optimization and changes to legacy code were needed to get it running. More-drastic changes were required for stage two, which required more memory per processor-core than was available on Stampede 2. To mitigate this problem the memory intensive decay time histogram production was separated into a new third analysis stage. This reduced memory usage to acceptable levels, with the added benefit of enabling fast turnaround on minor adjustments or exploratory analyses that may now be performed with the stage-three analysis without re-running stage two.

The muon-capture rate is determined by fitting the decay time histograms produced by the stage-three analysis to an exponential decay. Because of the fusion-interference and electron-background effects immediately after the muon entrance, the decay curves have been fitted beginning after a one-microsecond delay. Combined with data-quality cuts this results in total statistics of 7.9×10^9 muon-electron pairs. The initial high-statistics lifetime fits show no unexpected irregularities, as seen in the example fit on the left side of Fig. 4.3-1.

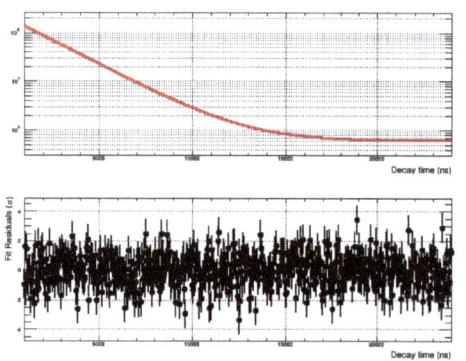 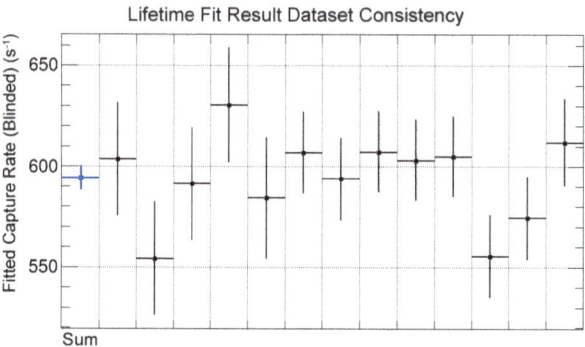

Figure 4.3-1. *Left:* Exponential fit of high-statistics decay curve from R2015 data pass (top), with corresponding fit residuals (bottom). The fit matches the data well, with no major irregularities in the residuals. *Right:* Extracted blinded capture rate for the summed R2015 data and for individual datasets throughout the run. The datasets have no obvious trend and exhibit good agreement with the sum, reflecting the steady and optimized running conditions during the 2015 production.

Due to the optimized conditions during the 2015 production, the data are quite consistent over time. This can be seen in the right plot of Fig. 4.3-1, where individual datasets show good agreement even at this preliminary stage. Applying more-sophisticated analyses such as the fusion-interference correction explained in (Sec. 4.2) should correct for minor changes in run conditions and bring these datasets into even-closer alignment.

In addition to the μ^- data, 1×10^9 μ^+ events were processed. Positive muons are useful for investigating systematics, since they do not undergo many of the interactions that negative muons do. In particular, these data are being analyzed to study the neutron detector backgrounds. Because positive muons neither capture nor cause muon-catalyzed fusion, they should not produce any neutrons. However, neutrons are still observed in the detectors, due to a process whereby a decay electron scattering in the wall materials can knock out a so-called "photo-neutron". These neutrons are a significant background for neutron-capture studies; the μ^+ data allow us to quantify this background.

4.4 Measurements with the MuSun electron tracker

D. W. Hertzog, P. Kammel, E. T. Muldoon, D. J. Prindle, and D. J. Salvat

Extracting the disappearance rate to 10 ppm demands a rigorous understanding of the decay electron detector stability and sources of background. Preliminary fits over the 2014 and 2015 data-taking campaigns using the "eSC", which is a segmented scintillator array, give good agreement over a range of different experimental conditions. The inclusion of two concentric cylindrical wire-chambers ("ePC1" & "ePC2") reduces backgrounds by more than an order of magnitude (see Fig. 4.1-1). This gives the potential to suppress the muon beam-related backgrounds studied extensively in the 2016 study of systematic effects at PSI[1]. However, an initial analysis of full tracks using information from both the wire-chambers and scintillator array did not produce results consistent with the analysis using only the scintillators. This motivated an effort to study the wire-chambers in greater detail, leading to numerous improvements to the MuSun electron detector analysis framework over the past year.

Pulses in the wire-chamber detectors ("wire hits") are spatially and temporally clustered into events at an early stage of the MuSun analysis. These events are later analyzed to identify coincidences among the two concentric wire-chambers and the scintillators. The number and distribution of wire hits comprising a cluster can provide a useful diagnostic check of the detector performance. We recently incorporated information about the number and multiplicity of these clusters through later stages of the analysis, permitting a systematic comparison of the fitted disappearance rate for varying detector conditions. In addition, the scintillator and wire-chamber event reconstruction algorithms were rewritten into a more concise object-oriented framework, improving upon the analysis used for earlier commissioning runs of the experiment.

Along with improvements to the analysis, the various electron track topologies were detailed using graphical event displays. This permitted finer tuning of the timing and angular acceptance cuts used to construct full tracks, and revealed sources of instrumental noise and combinatorial backgrounds which were mitigated by the improvements to the analysis. Fig. 4.4-1 shows an example of a full track with multiple-hit clusters in the wire-chambers, and thus multiple combinations of potential electron tracks. Events such as these were examined, with a strict set of criteria to choose one principal track to include in the analysis and avoid complications due to detector noise and backgrounds.

---

[1]CENPA Annual Report, University of Washington (2017) p. 73.

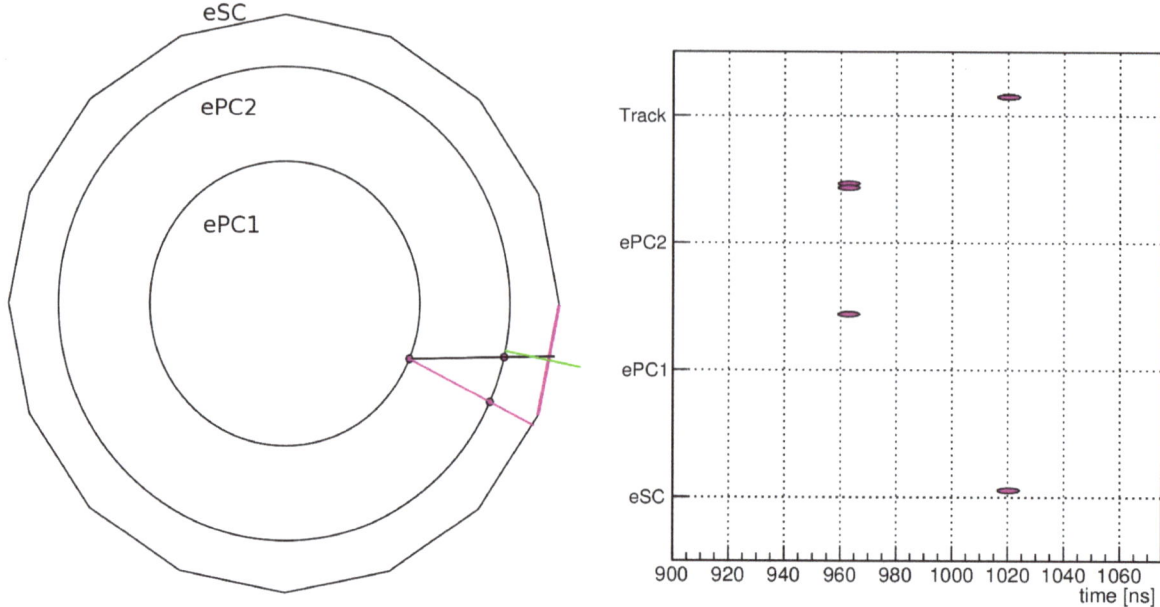

Figure 4.4-1. An electron-decay event with multiple potential electron tracks. *Left:* A schematic view of electron detector events, viewed along the beam axis. The two concentric circles represent the wire-chambers, with magenta dots representing clusters of wire hits. The outer hexakaidecagon represents each of the sixteen double-layer scintillators; segments with four-fold coincidences are colored in magenta. Each line passing through the three detectors represents a potential track, with the black line the uniquely-identified principal track, having the best spatial and temporal coincidence. *Right:* The wire-chamber and scintillator events from the left panel, plotted versus time, with ePC1(2) the inner (outer) wire-chamber, eSC the double layered scintillator segment, and "track time" corresponding to the coincident scintillator time.

Fig. 4.4-2 shows a comparison of two typical electron definitions in MuSun. The normalized residuals to a high-quality portion of 2014 production data are in reasonable agreement. Further studies are underway to study the instability of the wire-chambers relative to the well-behaved scintillator events for the 2014 and 2015 production data. Beam-related studies in 2016 will be used to further study any potential systematic effects, including a dataset wherein the electrostatic kicker was disabled. Further studies of the diagnostic information from the electron wire-chambers are also underway.

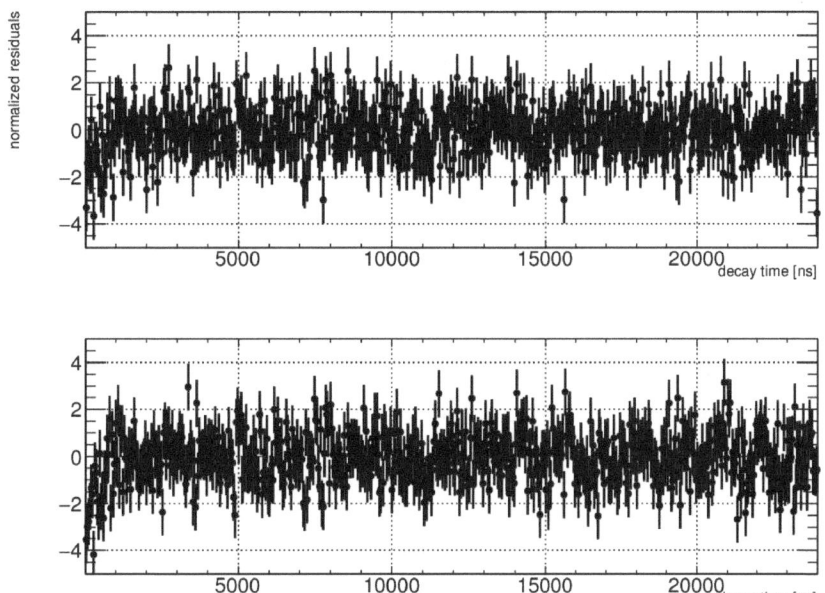

Figure 4.4-2. Fit residuals for two different electron track definitions used in MuSun, for a high-quality dataset from the 2014 production campaign. *Top:* The combined scintillator and wire-chamber electron tracks, with an analysis cut requiring that the electron track originates from the location of the muon stop in the TPC. *Bottom:* Electron events requiring only the scintillator array. Both electron definitions are consistent with the expected exponential behavior, but the fitted lifetimes for these definitions are discrepant in earlier datasets, the cause of which is currently under investigation.

4.5 Monte Carlo framework and studies

D. W. Hertzog, P. Kammel, E. T. Muldoon, D. J. Prindle, R. A. Ryan, and D. J. Salvat

The MuSun Monte Carlo simulation includes detailed descriptions of the μ^- kinetics as well as the time projection chamber (TPC) response. The rest of the MuSun hardware including muon entrance counters, electron detectors, and neutron detectors are described realistically in GEANT but have a nearly ideal detector response. We analyze the Monte Carlo data in parallel with real data but, knowing the input capture rate, we do not blind the Monte Carlo. The statistical uncertainty on the fitted rate of our current Monte Carlo electron sample is less than 10 s^{-1} but our decay rate fits were consistently 60 s^{-1} lower than our input capture rate. This discrepancy was tracked down to the physics implementation in GEANT which used an outdated and incorrect theoretical reference for the modification of the free muon decay rate in a muonic atom. We fixed that and notified the GEANT team of the more accurate reference[1] which gives a bound state correction of 12.1 s^{-1} for MuSun.

The TPC is segmented into a 6 by 8 grid of pads, and several tracking algorithms have been developed to reconstruct the muon stop position from the pad energy depositions[2].

---

[1] H. Überall, Phys. Rev. **119**, 1 (1960).
[2] CENPA Annual Report, University of Washington (2015) p. 89.

Muon catalyzed fusion events produce either a proton and a triton or a neutron and a ^{3}He, which deposit additional energy in the TPC and may interfere with the trackers. Proton-triton fusion, in particular, has a large effect on our tracking algorithms, since the protons are produced with enough range to cross into adjacent detector pads. The resulting position mis-reconstruction has been studied extensively, but now we have also studied whether fusions can cause a loss in detector efficiency.

The threshold tracker assumes the muon stopped on the first pad row to exceed an energy of 1.1 MeV. It has very high efficiency for both proton-triton and fusion-less events, although sometimes the stop position is reconstructed incorrectly. Alternatively, the upstream tracker attempts to fit the track pad energies to a μ^- Bragg curve. This fit will fail if energy deposition from fusion products is included, so the last two pads of the track are omitted from the fit to avoid fusion pulses. The upstream tracker was about 0.2% inefficient for events with proton-triton fusions when the μ^- stopped near the center of a pad. This was due to events with two proton-triton fusions, in which one proton traveled downstream and the other went upstream. Such a configuration deposits energy on the last three pads of the track instead of the last two, again causing the upstream fit to fail. We solve this problem by also omitting the third pad from the fit when its energy exceeds a threshold of 1 MeV.

There are also a small fraction of n-^{3}He fusion events in which the neutron scatters off a deuteron and the recoil energy is added to the track. The threshold tracker can fail if the extra recoil energy pushes an early pad above the threshold, causing about 0.05% inefficiency from fusions at the front of the fiducial volume. The impact on lifetime fits is likely to be small, but must be quantified. The recoil energy can also cause the upstream tracker Bragg curve fit to fail. We have implemented tighter time coincidence cuts when including pulses in the upstream track, reducing this effect to negligible levels.

Neutrons from n-^{3}He fusions can also interact in the plastic electron scintillator (eSC). This affected the lifetimes extracted from fits of the eSC time distribution. The recoils in the plastic scintillator are protons or ^{12}C, both of which quench, producing of order 10% of their deposited energy in the form of light. After we introduced a quenching model and a realistic threshold into the eSC response the decay rate shift due to n-^{3}He fusions was reduced to negligible levels.

Muon $g-2$

4.6 Overview of the Muon $g-2$ experiment

H. Binney, M. Fertl, A. T. Fienberg, N. S. Froemming, A. García, J. B. Hempstead, D. W. Hertzog, P. Kammel, J. Kaspar, K. S. Khaw, B. MacCoy, R. E. Osofsky, and H. E. Swanson

Context

Muon $g-2$ is a special quantity because it can be both measured and predicted to sub-ppm precision, enabling the so-called $g-2$ test for new physics defined by $a_\mu^{\text{New}} \equiv a_\mu^{\text{Exp}} - a_\mu^{\text{SM}}$. As a flavor- and CP-conserving, chirality-flipping, and loop-induced quantity, a_μ is especially sensitive to new physics contributions[1]. We update the $g-2$ test compared to our previous Annual Reports owing to a 2018 evaluation of hadronic vacuum polarization (HVP) that takes into account not only all recent experiments, but also a comprehensive correlations analysis.[2]

$$\Delta a_\mu^{\text{New}} = [270.5 \pm 72.6] \times 10^{-11} \quad 3.7\,\sigma. \tag{1}$$

The persistent discrepancy between experiment and theory fuels speculative models of new physics; however, tension exists with respect to many of the most popular interpretations owing to continually improved supersymmetry (SUSY) limits from the LHC at high-mass, and strict limits on dark photons at low-mass. Still, there is room for explanations in other model extensions if the discrepancy is confirmed. Our UW group has been helping to lead a next-generation $g-2$ experiment (E989) at Fermilab, which aims to improve the BNL E821 final result[3] by a factor of 4. Our final goal is a relative precision of 140 ppb on a_μ. In the fourteen years that have passed since the BNL final result was published, the Standard Model (SM) uncertainty has been greatly improved. Anticipated theory improvements on the timescale of E989 data-taking aim to match the uncertainty goal of the experiment. In Fig. 4.6-1, the more-recent SM evaluations are displayed with respect to the BNL result and the anticipated final precision of the Fermilab experiment.

---

[1] A. Czarnecki and W. J. Marciano,"The Muon anomalous magnetic moment: A Harbinger for new physics,"Phys. Rev. D **64**, 013014 (2001); D. Stockinger, "The Muon Magnetic Moment and Supersymmetry," J. Phys. G **34**, R45 (2007).

[2] A. Keshavarzi, D. Nomura and T. Teubner, "The muon $g-2$ and $\alpha(M_Z^2)$: a new data-based analysis," arXiv:1802.02995 (2018).

[3] Muon g-2 Collaboration: G.W. Bennett *et al.*, Phys. Rev. D **73** 072003, (2006).

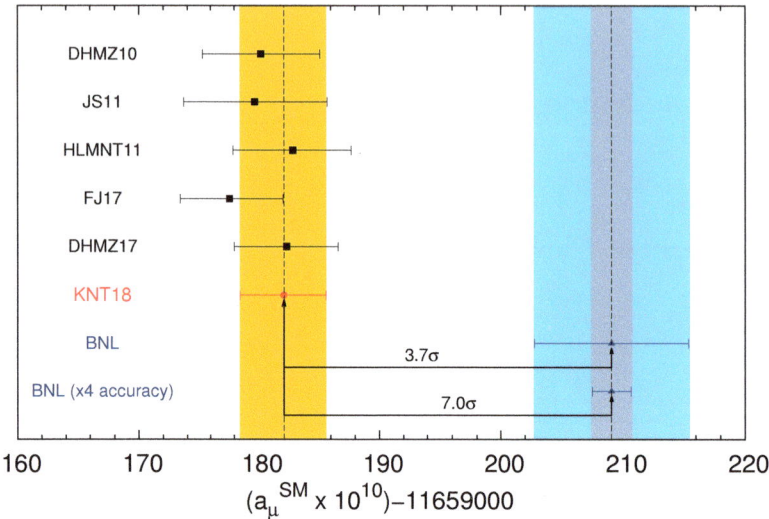

Figure 4.6-1. Comparison of Experiment to Theory for a recent history of SM evaluations based on treatment of leading-order hadronic vacuum polarization, which has the largest uncertainty. The yellow band represents the current average value and uncertainty. The blue band represents the E821 experiment, published in 2004. The gray inner band represents the expected precision of the new experiment; it's placement is simply centered on the previous experimental result. The differences, in standard deviations, are indicated. Figure from Keshavarzi et al, with their 2018 evaluation indicated in red by KNT18.

The SM terms are usually listed in five categories:

$$a_\mu^{\text{SM}} = a_\mu^{\text{QED}} + a_\mu^{\text{Weak}} + a_\mu^{\text{HVP}} + a_\mu^{\text{Had-HO}} + a_\mu^{\text{HLbL}}. \qquad (2)$$

The QED, Weak, and hadronic higher-order (Had-HO) terms have negligible uncertainties. The HVP precision continues to improve; it can be determined from experiment through a dispersion relation that amounts to an energy-weighted integral of $e^+e^- \to hadron$ total cross sections. The hadronic light-by-light (HLbL) effect has been evaluated using models and the quoted uncertainty of 26×10^{-11} is only a consensus estimate reached by comparing a variety of models; it is not a well-defined uncertainty. What is promising is the rapid progress being made to calculate the HLbL contribution to high precision using the lattice, a subject on which we commented in last year's Annual Report. This year, we mention a new data-driven approach to HLbL that has been in development for some years and is now bearing fruit. Hoferichter *et al.* calculated the leading pion-pole contribution with pure data-based inputs, including uncertainties.[1] They confirm previous model-driven methods and establish this leading term with higher precision and with uncertainties that can be quoted in a final SM evaluation.

---

[1] M. Hoferichter, B. L. Hoid, B. Kubis, S. Leupold and S. P. Schneider, "Pion-pole contribution to hadronic light-by-light scattering in the anomalous magnetic moment of the muon," arXiv:1805.01471 (2018).

Experiment and Collaboration

The muon anomaly is proportional to the ratio ω_a/ω_p, where ω_a is the anomalous precession frequency of the muon spin in a magnetic field and ω_p is a measure of that average magnetic field determined using proton nuclear magnetic resonance (NMR). Both frequency measurements must control systematic uncertainties to 70 ppb. The statistics required to determine ω_a exceed those at Brookhaven by a factor of 20. Our UW group is involved in both the ω_a and the ω_p measurements. We are designing and building a significant array of hardware tools in both cases. Additionally, we have modeled the optimization of muon storage in the ring, including tuning strategies for the superconducting inflector and the kicker field strengths, and the quadrupole and collimator geometries. A set of entrance imaging counters and a luminosity monitor were designed and installed to aid in the tuning phase and in the evaluation of beam intensity.

The most important update we can provide in this Annual Report is that the Muon $g-2$ experiment is now running and collecting physics data. Over the past year, the we completed an extensive beamline and experiment commissioning campaign and have begun to acquire meaningful physics data. As of mid-May, 2018, we have accumulated a dataset slightly exceeding the statistics obtained throughout the entire BNL E821 experiment. However, there remains a long way to go; this benchmark — while important — represents just 5% of our eventual statistics goal. The status of the experiment is described in (Sec. 4.7).

The E989 Collaboration has 35 institutions and > 160 collaborators from 8 countries. D. Hertzog is completing his last term as Co-Spokesperson, having helped found and lead the experiment since 2009. J. Kaspar acted as co-Run-Coordinator this year, thereby helping steer the day by day operation of the experiment. E. Swanson has been Deputy Field Team Leader. K. S. Khaw is co-Analysis Coordinator for the precession measurement. A. Fienberg continues to author and manage the Data Quality Monitor (online) information. J. Hempstead is the Calorimeter coordinator. Graduate student M. Smith completed his Ph.D. work on the precision magnetic field determination and shimming. Our group currently has 6 additional Ph.D. students with two expected completions in the next 6 months. The collaboration's construction work is complete. Technical Teams that are responsible for the work are generically named: Beam, Ring, Field, Detector, Simulation, and Offline. Our group is involved in almost all of these teams at various levels.

The Beam delivery design involves, in order, the Booster, Recycler, Transfer Lines, Target Station, Decay Beamline, Delivery Ring (DR), and final M4/M5 beamline. This suite is necessary to create and deliver a bunched, 3.1 GeV/c polarized muon beam, which is purified of background pions and protons by running the bunch around the DR for several turns prior to injection in our storage ring. This effort is led by Fermilab Accelerator Division (AD) with collaboration members providing some of the modeling. The commissioning of the new beamlines has taken a number of months during this past year and the delivered muon flux is now nearly at the level anticipated in the Technical Design Report. Graduate student Froemming played a leading role as an interface between the AD group and the Collaboration and spent considerable time personally optimizing the beam tunes into the storage ring; see (Sec. 4.8).

The Ring team is responsible for building and operating the storage ring. They also provide the inflector, quadrupoles, collimator, and kicker subsystems. The team carries out simulations of muon storage and evaluates beam-dynamic systematic uncertainties. We designed and built a set of imaging detectors that are positioned along the lengthy magnet-free corridor between the last beamline quadrupole and the entrance to the inflector magnet. They are needed to aid beam tuning as they provide images of the beam in real time that are required to optimize muon transmission into the ring. P. Kammel and graduate student B. MacCoy describe the work in (Sec. 4.13). To monitor the incoming beam intensity and time profile, graduate student H. Binney prepared a $T0$ detector system, which is positioned immediately downstream of the final beamline quadrupole. She describes the status in (Sec. 4.9).

The Field team completed coarse shimming of the magnet by November, 2016. Access to the magnet was then put on hold until the vacuum chamber system and the rest of the experimental equipment was installed, many months later. During this period, we installed and tested the UW-built Fixed Probe NMR system, which includes 378 NMR probes located in special grooves in the vacuum chambers and all of the pulsing and acquisition electronics. The system is described in (Sec. 4.14) by E. Swanson. Next, fine-shimming was completed and the azimuthally-averaged multipole distribution was obtained from data acquired using the in-vacuum NMR Trolley system. The final step in field preparation is minimization of the residual multipole components. This is achieved by employing 400 concentric current loops that circumnavigate the ring above and below the vacuum chamber. Each can carry a unique current, as provided by individual power supplies. The task of testing and debugging this Surface Coil system was carried out by graduate student R. Osofsky. She also modeled patterns of coil currents that would produce various multipole or radial fields and, most importantly, the final current settings that minimized the multipoles. The work is described in (Sec. 4.15).

The Detector Team provides the instrumentation needed to monitor the stored muon distribution and to measure the decay positrons from which the characteristic precession signal histograms are built. The UW group designed and built the electromagnetic calorimeter system, which consists of 24 stations, each having arrays of 54 PbF_2 crystals with large-area SiPM readouts. The electronics, testing, and mechanical supports were all developed at CENPA. All 1296 crystals are now fully calibrated and functioning well during this first commissioning run. However, this was not quite the case when the intense beam first arrived at the ring. The initial intense beam splash overwhelmed the electronics and sagged the gain of our detectors. With a rapid setup to reproduce the effect here at CENPA, and insight from our electronics personnel, a fix was quickly conceived, built and executed, prior to the beginning of the physics data-taking period. Graduate student J. Hempstead led the initial calorimeter installation and commissioning work, and then led the testing and mitigation program to stabilize the detector gains. He describes the work in (Sec. 4.10)

Last year, graduate student A. Fienberg and postdoc K.S. Khaw began developing the offline analysis framework for the experiment. Their work is now quite mature. It involves interpretation of the raw data as fitted-pulses with calibrated times and energies, shower crystal-clustering, muon-loss identification, pileup-subtraction algorithms, histogram

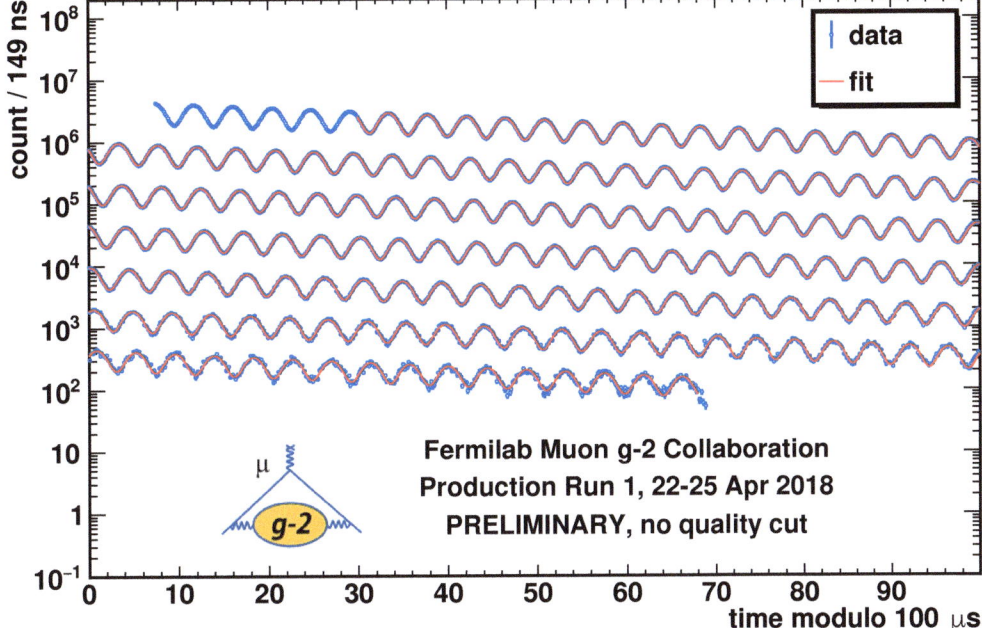

Figure 4.6-2. A typical muon spin precession frequency plot from a short period of data accumulated in April 2018. The data points are the number of positrons above 1.8 GeV from muon decay vs. time after muon injection into the storage ring. The clearly evident oscillation encodes the orientation of the muon spin. At the peaks, the muon spin points in the direction of motion; at the troughs, the spin direction is reversed.

construction, and sophisticated fitting toward extraction of the precession frequency. They have both built online tools based on the offline framework that are used continuously by on-shift monitors to evaluate data quality and to help experts in beam tuning exercises. They describe aspects of their recent work in (Sec. 4.11) and (Sec. 4.12). To close this Overview, we provide a snapshot of a fitted precession spectrum based on a recent 60-hour period of running in Fig. 4.6-2. The preliminary analysis of these data is encouraging.

4.7 Status of Muon $g-2$ experiment

D. Hertzog and J. Kaspar

The first commissioning run of the Muon $g-2$ experiment began in late spring 2017. It was followed by a planned accelerator shutdown over the summer, and then a startup of beam operations in late fall. The combined commissioning and first physics data-taking campaign began by December, 2017 and has run continuously, with an expected end of run on July 7, 2018. The transition to meaningful physics data-taking began in late March. The approximate timeline and main activities are sketched below.

Timeline

- Spring 2017 6-week Commissioning Run: In this period, infrequent injections of mixed protons, pions, and muons were delivered to the Storage Ring at less than 1 Hz. These data were used to do preliminary timing and modest muon-storage studies. During this period, we learned about various noise sources affecting the calorimeters, trackers, and kickers, which were subsequently mitigated in summer 2017.
- Shutdown period 2017: During these months, we completed key installation tasks, including the magnetic-field equipment, the magnetic-field trolley, the plunging-probe system, and the 16 straw-tracker modules. The DAQ was exercised nearly continuously toward toward the end of this period using the laser calibration system to artificially generate "data-like" events.
- Commissioning with beam, late-fall 2017 to early 2018: During this period, the proton / pion / muon beam was directed around the Delivery Ring to develop the process of proton and pion removal. Pions are removed by decay and protons, which travel at lower velocities, are removed by a fast kicker magnet. This period was also used for combined beam-injection tuning, detector optimization, and muon-storage parameter sweeps.
- February 2018: Lost to a vacuum incident during new cryopump installation. The vacuum breach caused damage to internal quadrupole plates, requiring repair.
- March 2018: This month was used to more-deeply understand muon-storage optimization and, thereby, the sources of apparent under-performance by the kicker. Nevertheless, we started physics production on March 22nd at 16 fills per 1.4 s accelerator cycle. At this time, our proxy for muons per fill — CTAGS — was reading 200, compared to an anticipated 1000.
- April 2018: We discovered that the kicker HV calibration was wrong and the set-point of 55 kV had not been reached. With re-calibrated parameters, the CTAG (CTAG is a value counting the number of high energy decay events reconstructed in the $g-2$ ring after a certain time, a value proportional to the number of stored muons.) rate increased to about 500 / fill, exceeding the rate achieved at Brookhaven National Lab (BNL). While this rate remains lower than planned, the sources of the next factor of 2 are a number of factors that are nearly performing to specifications, but not quite. These factors are slated for improvement in Summer Shutdown 2018 and have recently been reviewed in-depth by an internal FNAL committee for resource funding and support.
- May 11, 2018: The milestone of 1 x BNL total statistics was reached and we continued data taking in May and early June. At the time of this report, the accumulated statistics exceeds those from the BNL measurement and should lead to an important first publication.

Performance Highlights

While focus on acquiring physics data from optimized muon storage is paramount, it is worth reviewing the many systems (overseen by Jarek Kaspar as Run Coordinator) that have been successfully commissioned and are operating at, or above, design specifications. These include:

- Entrance T0 timing and intensity monitor (Sec. 4.9);
- The IBMS entrance X-Y profile imaging system (Sec. 4.13);
- The in-vacuum scintillating fiber "harp" system that images the stored muon beam (destructively, when rotated to measuring position);
- The in-vacuum, straw-tracker chamber system that is used to image the beam by tracing-backward the decay positrons.
- The UW calorimeters that acquire the central positron-decay data which are embedded with the precession frequency that is key to the measurement. They required an expedited gain-stability upgrade after beam was injected into the ring. This is described in (Sec. 4.10).
- The Italian laser calibration system that is used to establish the extraordinary gain stability requirements for the calorimeters and which is used to exercise the DAQ system.
- The Cornell 800 MSPS, 12-bit-depth deadtimeless digitizer system used to read out the calorimeters and other systems.
- The fast DAQ that can process the 20 GBytes/sec of raw data into a recordable steady stream at more than 200 MB/s.
- The online data quality monitor[1] and Nearline full analysis (Sec. 4.8).
- The offline processing using *art* framework, used for all data analysis. For examples of the work, see (Sec. 4.11) and (Sec. 4.12).
- The fixed probes (FP) (Sec. 4.14) NMR system that tracks the magnetic field 24/7. A subset of the FPs are used in a feedback loop to control the main magnet power supply to keep the average field constant.
- The trolley NMR system maps the field every few days, and the absolute-field plunging-probes system established the absolute magnetic field measurement.
- The final stage of shimming is made using active adjustments, including the 200 surface coils that are tuned to minimize azimuthally-averaged multipole moments in the field (Sec. 4.15).
- An interface was developed to exchange information between measurements made by the experiment and upstream magnet-tuning, as controlled by acceleration of the muon beam (Sec. 4.8).

Figure Fig. 4.7-1 shows three constantly-updated curves that provide a snapshot of the data taking. In blue are the total protons delivered to the experiment; in green the integrated DAQ uptime fraction, and in red, the raw number of positrons as a percentage of that achieved in the three years of running at BNL. At the time of this snapshot, we have accumulated about 1.5 times the total BNL statistics; however, about a third of those data will be unused owing to data quality cuts and various systematic error tests. We are actively working on the analysis of the "sweet spot" data period within this running period and plan to report the results sometime in 2019.

---

[1] CENPA Annual Report, University of Washington (2017) p. 89.

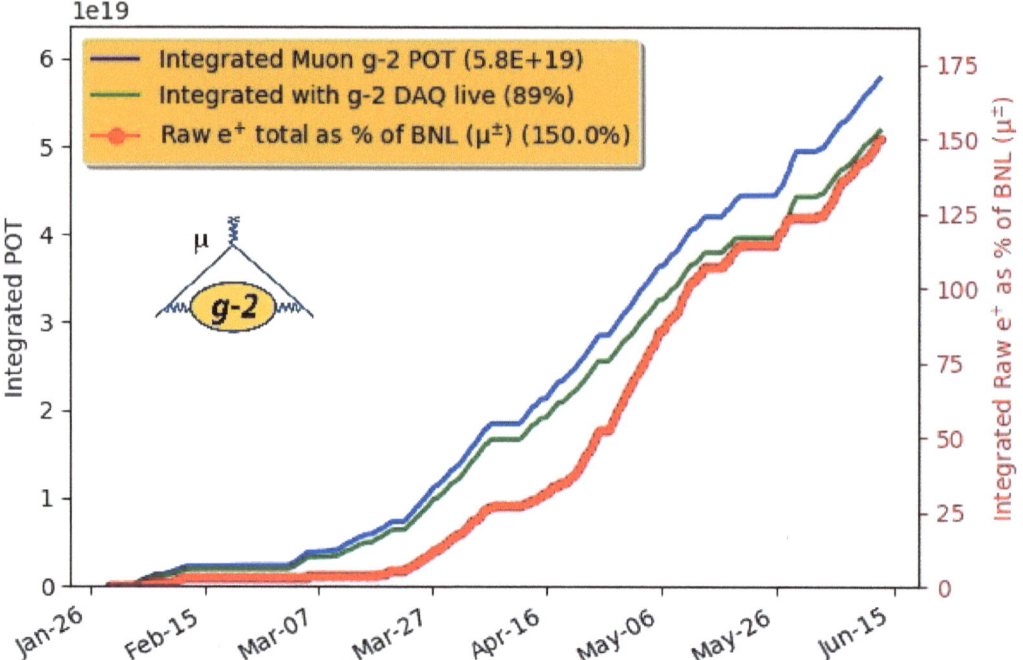

Figure 4.7-1. Plot showing, as of June 12, 2018, the total integrated positron-decay count compared to that achieved in three running periods at Brookhaven (red curve). From this total, we expect that data quality cuts and other systematic-test periods will eliminate the use of about a third of the data.

4.8 Muon campus M4/M5 beamline optimization, transmission and MULTs

N. S. Froemming

The muon $g-2$ experiment at Fermilab (E989) must collect 21 times more data than the previous BNL experiment (E821) in order to achieve the nominal goal of 100 ppb statistical uncertainty. Optimizing the properties of the injected muon beam into the storage ring plays a critical role in achieving this goal, and hence, in completing the experiment in a timely manner with the required statistics. A special superconducting septum magnet known as the "inflector" facilitates injection into the storage ring by providing a uniform dipole magnetic field that cancels the main dipole field of the ring, hence establishing a field-free corridor through which the injected muon beam passes as shown in Fig. 4.8-1. The unique-to-$g-2$ (E989 and E821) truncated-double-cosine-theta geometry of the discrete superconducting currents largely traps the return magnetic flux, and a superconducting shield further prevents any remnant fields from disturbing the high-precision dipole magnetic field required by the experiment. In this way, the muon beam is injected as close as possible to the nominal storage-orbit radius, $\rho_0 = 7.112$ m, while the perturbations to the highly uniform magnetic field in the muon storage region are kept to a minimum.

Stability of the injected muon beam's centroid motion and focus, through the narrow inflector aperture, is essential: Small horizontal-position offsets cause excessive scraping in the narrow inflector beam channel ($\Delta x = \pm 9\,\text{mm}$), while small horizontal-angular offsets ($\Delta x' \approx \pm 2.5\,\text{mrad}$) can mean the difference between optimized injection or storing no beam at all. As a result, small dipole corrector magnets known as "TRIMS" at Fermilab are used to steer and optimize the horizontal and vertical beam centroid through the inflector. A custom set of beam-tuning scripts written in Fermilab's Accelerator Controls Language (ACL) has been created in order to quickly and effectively optimize TRIMS settings (as well as other magnet settings) like the results shown in Fig. 4.8-2 (left). In general, experiment and simulation have been found to agree for a variety of magnet settings.

A conceptual overview of the procedure used to tune and optimize beam focusing through the inflector is shown in Fig. 4.8-2 (right). First, the horizontal and vertical beam parameters are measured using a quadrupole-scan technique about $20\,\text{m}$ upstream of the ring. Next, the measured beam parameters are propagated downstream to the end of the beamline (just upstream of the ring) to connect with simulations. Results from simulation are then used to determine where, in the landscape of optimal injection parameters, the measured beam parameters lie, and to formulate an effective tuning strategy from that point. Simulations indicate a highly correlated "ridge" of optimal injection parameters (β, α) in both transverse x- and y-directions, which reflects the need to focus the beam horizontally and vertically through the narrow transverse aperture of the 1.7 m-long inflector. The final six magnetic quadrupoles immediately upstream of the ring are then adjusted simultaneously using current ratios chosen to target specific features of the landscape. The concept of adjusting several beamline elements at once is known as a "MULT" at Fermilab; 8 quadrupole MULTS are constructed in total, i.e. 4 to target each of the canonical directions $\{\hat{\beta}_x, \hat{\alpha}_x, \hat{\beta}_y, \hat{\alpha}_y\}$, and another 4 to step "parallel" or "perpendicular" to the ridge of optimality shown in the landscape of Fig. 4.8-2 (lower-left). In this way, the abstract space of injection beam parameters is quickly and effectively explored, and the focusing of the beam into the $(g-2)_\mu$ storage ring is rapidly optimized (Fig. 4.8-2, lower-right). The observed variation of muon capture vs. injection beam parameters is found to agree well with simulation.

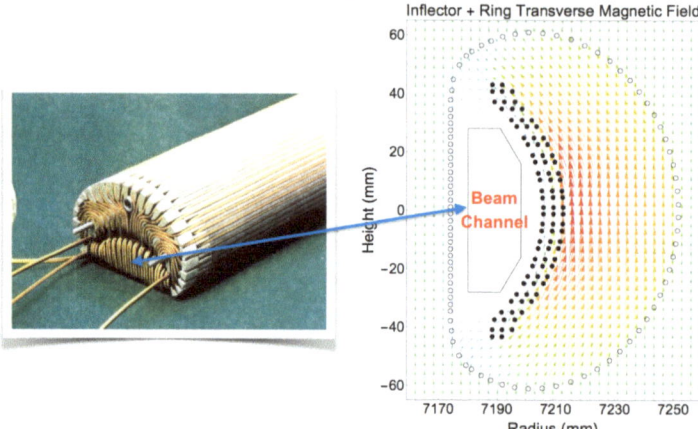

Figure 4.8-1. *Left:* Superconducting coils of the "inflector" magnet. *Right:* Vector sum of inflector and storage-ring transverse dipole magnetic fields. The discrete superconducting currents are arranged in a distinct truncated-double-cosine-theta pattern, and the resulting region of pure dipole field within the innermost coil is used to cancel the main dipole field of the muon storage ring. The return magnetic flux is trapped by the coil geometry, so the perturbations to the highly uniform storage-region field (right panel, left-hand side) are kept to a minimum. This allows the muon beam to be injected as close as possible to the nominal storage orbit, $\rho_0 = 7112\,\text{mm}$.

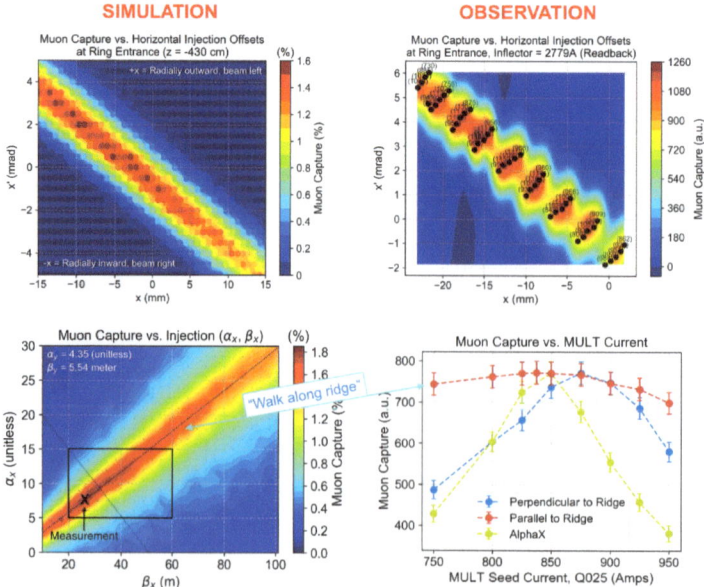

Figure 4.8-2. *Top:* Muon capture vs. horizontal position and angle (x, x') near the storage ring entrance according to simulation (*top-left*) and experiment (*top-right*). *Bottom:* Muon capture vs. horizontal beam-focusing parameters (β_x, α_x) near the storage ring entrance according to simulation (*bottom-left*) and experiment (*bottom-right*).

4.9 T0 detector, beam pulse shapes, and timing

H. Binney, T. H. Burritt, A. T. Fienberg, and D. W. Hertzog

The T0 start-time detector is the most-upstream detector in the magnetic storage ring. It measures the muon-beam intensity-versus-time profile as the muons are injected. It is located in the beamline just upstream of the IBMS 1 detector. This year, a new version of T0 was designed, built, and tested at CENPA. Previously, the detector was composed of three silicon photomultipliers (SiPMs) coupled to a thin plastic scintillator, through which the particles passed. Because of the high intensity period of primary beam injection, the SiPMs were often operating in a non-linear regime, causing distortions in the detector profiles. Therefore, we decided to replace two of the SiPMs with photomultiplier tubes (PMTs), which have a well-understood linear response, retaining the scintillator and the third SiPM. We tuned these two PMTs to be linearly sensitive to the main beam injection and tuned the remaining SiPM to be sensitive to low-intensity beam leakage during out-of-injection times. A hinged mechanism in front of each PMT allows insertion of neutral-density filters to control the amount of light entering the PMTs. An optical fiber, coupled to both channels, provides laser pulses that help to synchronize the T0 detector timing with the rest of the experiment. A diagram of the new T0 detector is seen in Fig. 4.9-1. Currently, one of the PMTs is a high photostatistics channel (lighter ND filter) and one is a low photostatistics channel (darker ND filter). A sample T0 trace from a single fill is seen in Fig. 4.9-2.

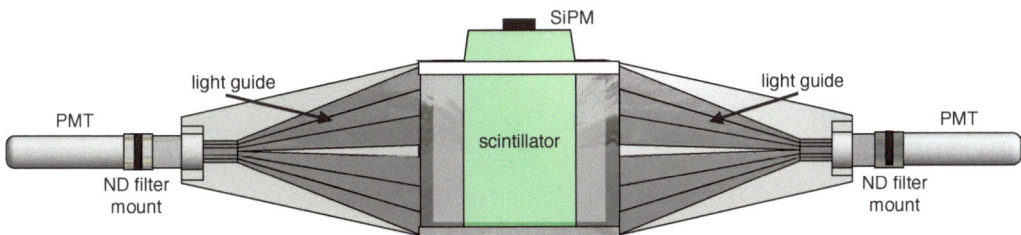

Figure 4.9-1. Diagram of the T0 detector. Particles entering the ring pass through the scintillator (in green). Light guides on either side of the scintillator transport the scintillation light to photomultiplier tubes (PMTs) on each end. Neutral density filters are mounted between the light guides and the PMTs in order limit the amount of light entering the PMTs, so that the PMTs can operate in a linearly sensitive region during beam injection. A SiPM is also coupled to the scintillator and is sensitive to low-intensity beam leakage during out-of-injection times.

At CENPA and at Fermilab, an absolute calibration of the detector was performed using a Sr-90 β source to mimic minimum ionizing particles (MIPs), allowing us to make an approximate estimate of the number of particles passing through the detector.

The T0 detector provides real-time feedback about the timing and size of the incoming pulses. The T0 integral - the sum of the integrals of the pulses in the two channels - is the main denominator used in tuning many of the downstream systems, allowing normalization to the amount of beam entering the ring. It is also used for upstream tuning. Further, the detector provides information about longer-term trends, such as average profiles of the beam pulses. The beam arrives in 8 distinct pulses, which were each expected to have a

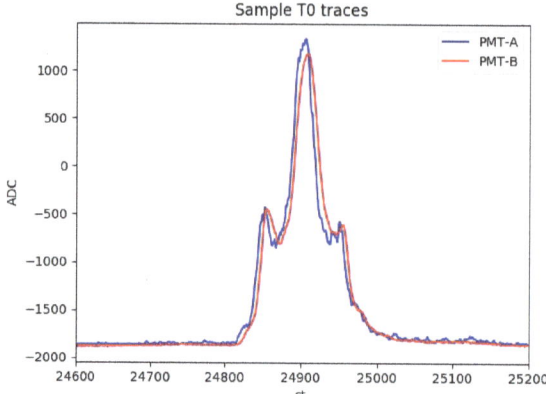

Figure 4.9-2. Sample T0 trace from a single fill showing the shape of a beam pulse passing through the detector. The horizontal axis represents time and is labeled in clock ticks (ct), where 1 ct $\approx$ 1.25 ns. The vertical axis represents intensity and is labeled in ADC counts, where 4 ADC $\approx$ 1 mV. T0 has two PMT channels, labeled as PMT-A and PMT-B in blue and red. These channels see the same signal from the scintillator; however, the PMT-A channel has more noise, due to lower photostatistics.

characteristic "W" shape. However, it was soon apparent that each of the 8 pulses had its own distinctive shape, which often did not resemble the anticipated "W". Using the T0 detector, averaged beam profiles were created for each of these pulses separately in order to help with modeling and simulation of the beam.

Reconstruction code was written for the T0 detector so that important T0 metrics would be available in the offline production data. Many of these metrics, including beam integral, beam timing, and beam RMS, are useful tools for data quality control. Events that have a very low integral value or have an abnormal arrival time or RMS are considered bad events and are excluded from the data used for analysis. A low integral value, for example, often indicates that no beam pulse has been delivered to the storage ring. Therefore, an event with a low integral value should be rejected. A data quality filter using the T0 detector has been implemented to remove these bad events.

4.10 Final installation of the calorimeters and gain sag mitigation

J. B. Hempstead, D. W. Hertzog, and J. Kaspar

Calorimeter hardware updates

During the summer 2017 shutdown at Fermilab, calorimeter problems that had been identified during the commissioning run were addressed. Most-pressing was the problem of noise that had limited the "island-chopping threshold". This threshold determines when a particular time-frame of digitized data is interesting and saves that information. Large noise spikes had prevented lowering the threshold, which also limited the minimum energy that could be reconstructed in the calorimeters. The main energy-calibration method of matching the minimum-ionizing particle (MIP) peak was hamstrung by the high threshold.

To reduce noise in the SiPMs, the grounding scheme was altered slightly. The output signal circuit of the SiPMs is no longer connected to the ground of the signal cables. Making this change for all 54 SiPMs in each of the 24 calorimeters was a substantial effort but allowed us to reduce the island-chopping threshold enough to calibrate the SiPMs using the MIP peaks.

In addition to the signal-cable changes, a number of channels had problems that required attention. Approximately ten HDMI cables required replacement due to problems communicating with the SiPMs. The cause of the failures is not understood, but new cables restored communication. Approximately five of the crystal/SiPM pairs (out of 1296) required replacement because those SiPMs blocked communication with all of the SiPMs attached to the same breakout board and the crystals and SiPMs are semi-permanently bonded.

The ±5,V supply cables for the SiPM-communication breakout boards were all remade. The original cables were unwieldy and could be electrically shorted by adjustment alone. The new cables were built under the careful supervision of CENPA electrical engineer David Peterson. No problems have been found with the new cables.

Continuing the in spirit of renovation, a significant effort was undertaken with David Peterson to improve the power-supply boxes. The design includes a logic circuit that can inhibit the output of the supplies if sent a command by the SiPM-controlling BeagleBones. During initial installation, this logic was bypassed in roughly 75% of the supplies. After investigation, David was able to re-establish the remote-control capability to many of the calorimeters' power supplies. $\sim 25\%$ of the calorimeters now have the logic circuit bypassed.

Gain sag mitigation

During the commissioning leading up to the first data-taking run for $g-2$, laser tests were performed by our Italian colleagues to probe the gain within the measurement period. Laser pulses were fired into all channels of the calorimeters while muon data were being taken. The size of the laser pulses "in fill" were then compared to a reference value to determine the gain stability of the system. Surprisingly, the gain was found to "sag" almost 8% immediately after muon injection into the storage ring. For reference, the design goal of the experiment is to have (corrected) gain stability of better than 0.04%.

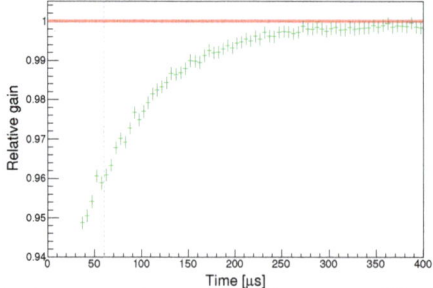

Figure 4.10-1. Relative calorimeter gain vs. time in muon fill. The size of laser pulses fired during muon storage is compared to a common reference point, shown at $t = 5000\,\text{ns}$. The red line indicates a perfectly stable gain, with the shaded region showing the stated stability goal. Muons are injected at $t = 30\,\mu\text{s}$ on this plot. The vertical dashed line indicates the start time of fits to extract ω_a.

Following this measurement, a full replica electronics chain was assembled at CENPA in order to perform tests and find a solution. An LED pulse was used to simulate the large "splash" of particles that the calorimeters see at injection, and a second LED was used to simulate the decay structure of a normal measurement period. Testing was performed with the guidance of CENPA electrical engineers David Peterson and Tim Van Wechel. After testing David Peterson designed a new feedthrough box, dubbed the "MegaBox", that holds 8 large capacitors for fast replenishment of charge to the SiPMs. Results from the test stand before and after the addition of the MegaBox can be seen in Fig. 4.10-2.

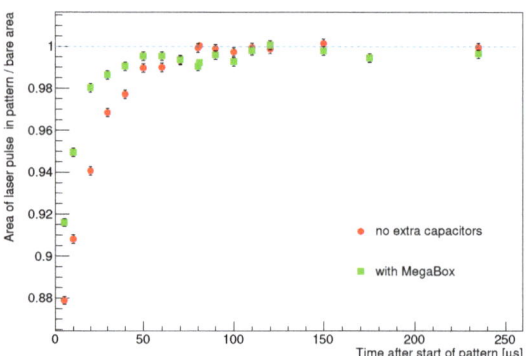

Figure 4.10-2. Relative pulse area vs. time in pattern. The ratio of the area of laser pulses recorded by a sample SiPM during the LED pattern to the area of the same laser firing without the LED pulse are plotted against time in the LED pattern. The LED pattern was designed to simulate the structure a calorimeter might see during a measurement period. The red circles are measurements taken before addition of the MegaBox, while the green squares are measurements taken with a MegaBox installed. The relative improvement at the beginning of the pattern and the improved recovery time motivated the installation of the MegaBoxes at Fermilab.

Twenty-four of the new MegaBoxes were assembled at CENPA and shipped to Fermilab for installation. Tests performed by the laser team showed the improvement that these boxes made, shown in Fig. 4.10-3.

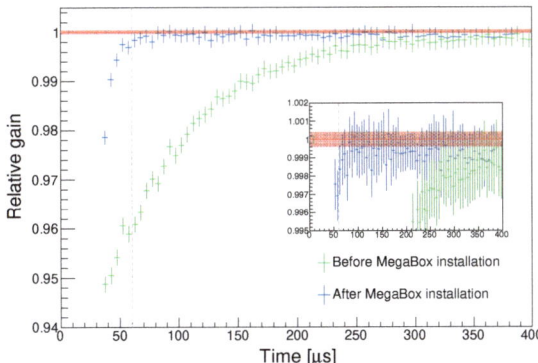

Figure 4.10-3. Relative calorimeter gain vs. time in muon fill. The energy of laser pulses fired during muon data collection and normalized to a common reference point, shown at $t = 5000$ ns, are plotted. Ideally, all the points would lie on the red line at relative gain = 1. The goal stability is shaded in red. Muons are injected at $t = 30\,\mu$s, and the gray dashed line is used to show the time used as the start time for fits. Data from before the hardware improvement are shown in green while data from after the fix are shown in blue. An inset is provided to show small-scale features near the performance goal. The improvement from the addition of the MegaBox is clear: both the initial drop and the recovery time are reduced.

4.11 Energy-binned muon precession analysis

A. T. Fienberg

In the laboratory reference frame, the muon-decay positron-energy distribution depends on the muon's longitudinal spin polarization. As the stored-muon population's spin precesses relative to its momentum, the observed decay positron energy distribution changes in time periodically at the anomalous precession frequency. In the E989 Muon $g-2$ experiment, the precession frequency measurement is accomplished by observing the time modulation of the positron energy distribution (Fig. 4.11-1).

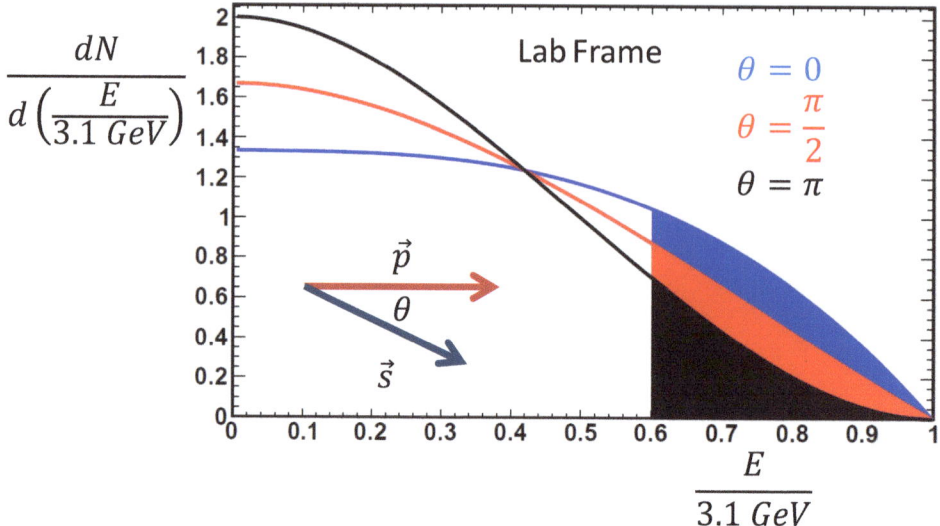

Figure 4.11-1. Theoretical muon-decay positron lab-frame energy distribution for differing longitudinal polarizations. The shaded region indicates the energies above the T method threshold described in the text. As the polarization rotates, the probability that a decay positron will be above this energy threshold oscillates.

The traditional analysis technique—called the T method—histograms the calorimeter hit times of positrons detected above a certain energy threshold. As the energy distribution changes with the muon population's polarization, the probability that a given decay positron will be over-threshold changes as well. Thus, the anomalous precession frequency is apparent in the high-energy positron hit-time histogram. The probability that a muon will decay in general cannot depend on its polarization, so the number of low energy positron decays oscillates out of phase with the number of high-energy positron decays. Thus, there is an optimal, non-zero energy threshold for the best statistical precision achievable by such a technique.

While the baseline analysis technique is sufficient to achieve the target statistical precision of the E989 Muon $g-2$ experiment, other techniques provide improved statistical power. Additionally, alternative analysis techniques generally have different susceptibilities to the various systematic effects present in the $g-2$ experiment. Consistency between differing techniques builds confidence in a final result.

One alternative analysis technique is an energy-binned analysis. Rather than forming a single positron hit time histogram, an energy-binned analysis forms a family of non-overlapping hit time histograms each containing the positrons detected in an energy range from E to $E+\Delta E$. As each positron comes from a separate muon decay, the non-overlapping hit time histograms are statistically independent. Each of these statistically-independent histograms is fit independently for the muon precession frequency. The resulting frequencies are averaged—with appropriate statistical weights—to obtain the final precession frequency result. An energy-binned analysis uses more of the data than the traditional approach and, through the averaging procedure, intrinsically weights the different energy positrons based on the power of their individual precession frequency measurements. The resulting precession frequency measurement is 10-15% more precise than the traditional analysis approach. The exact value of the improvement depends upon detector effects that must be measured.

Fig. 4.11-2 shows the results of applying this technique to a small dataset acquired in February 2018. The energy-binned analysis achieved 12% smaller uncertainty than the T method analysis.

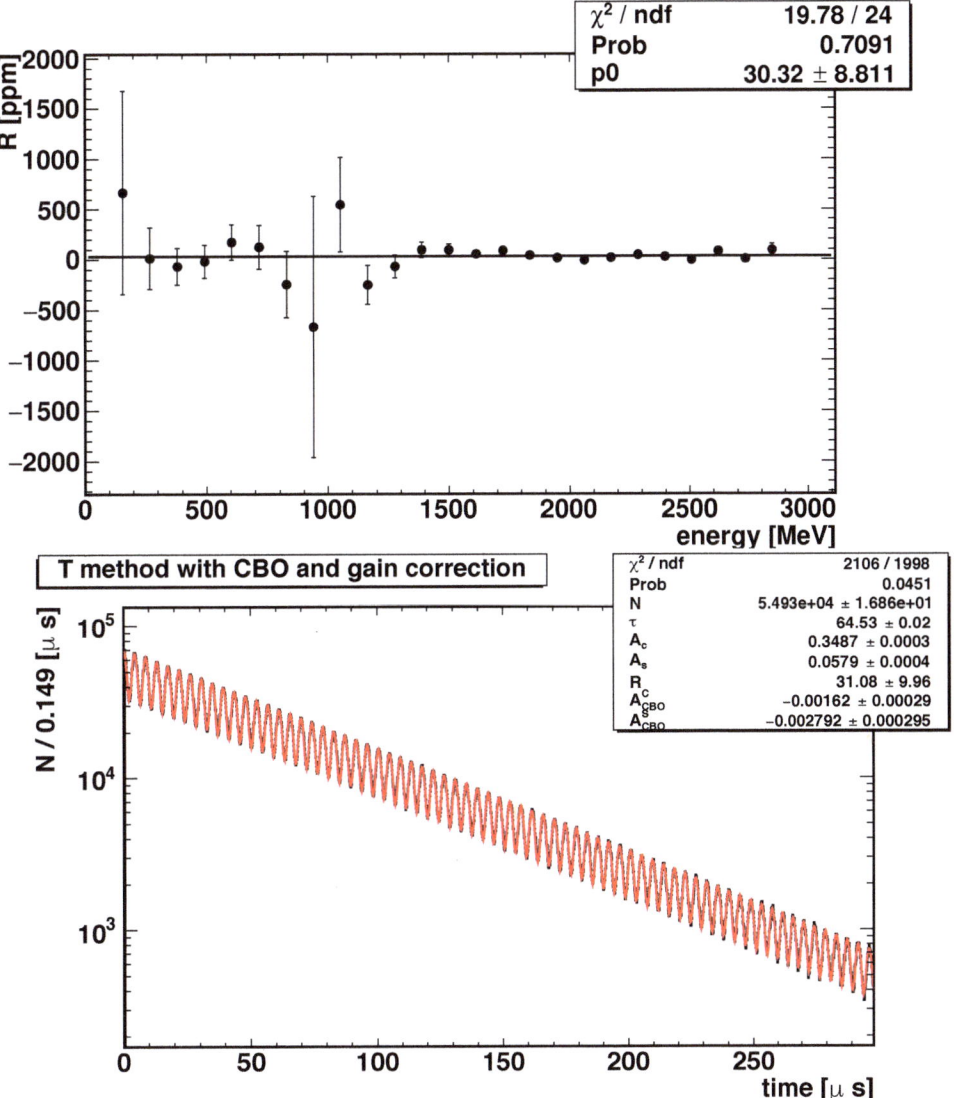

Figure 4.11-2. Energy-binned and T-method precession frequency analyses applied to data collected in February 2018. The R-parameter (p_0 for the energy binned fit) is the difference between the fitted frequency and a blinded reference frequency, in units of parts-per-million [ppm]. The energy-binned analysis achieves better precision, as evidenced by the smaller uncertainty. The low p-value, high χ^2 of the T method fit indicates that there may remain un-accounted-for systematic effects.

Rate-dependent calorimeter gain changes are a particularly challenging detector-based source of systematic uncertainty. As the phase of the observed oscillation varies with the detected positron energy, drifting calorimeter gains impart a time dependent phase shift that, if uncorrected, biases the extracted frequency. A Monte Carlo simulation was developed to test the size of this bias for differing analysis techniques. The simulation indicates that an energy-binned analysis on a dataset perturbed with rate-dependent gain shifts suffers a smaller bias than a traditional T method approach. The smaller bias is attributable to partial cancellation between biases in high- and low-energy bin fits (Fig. 4.11-3). The CENPA-based University of Washington analysis team will conduct an energy-binned analysis of the 2018 $g-2$ dataset.

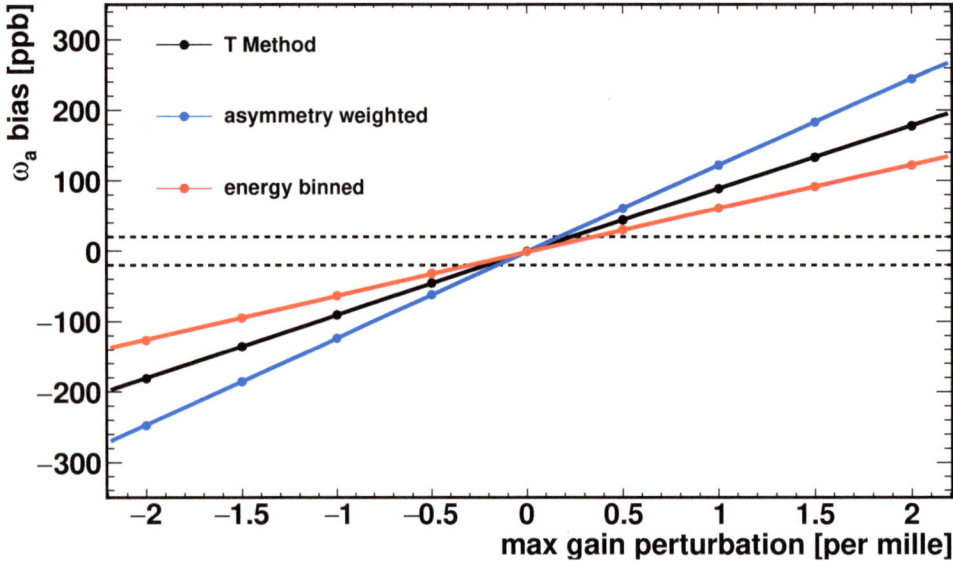

Figure 4.11-3. Result of a Monte Carlo simulation used to characterize the susceptibilities of varying analysis techniques to rate-dependent calorimeter gains. The asymmetry-weighted analysis is a variant of the T method analysis in which positron-energy-dependent weights are applied to the hit-time histogram. It has similar statistical power to the energy-binned analysis, but greater susceptibility to changing gains.

4.12 Performance of the calorimeter: timing and energy

K. S. Khaw and A. T. Fienberg

Lead fluoride calorimeters are an essential part of the muon $g-2$ experiment for measuring the muon anomalous-precession frequency. One of the challenges specific to the experiment is the energy calibration, from fitted-pulse integrals to particle energies in MeV. Unlike low-energy physics experiments where well-calibrated radioactive sources can be used or high-energy physics experiments where known resonances like $J/\psi \to \gamma\gamma$ or $Z \to e^-e^+$ can be utilized, there is no standard source for the muon $g-2$ experiment. The previous muon $g-2$ experiment at BNL empirically fixed endpoint of the observed positron energy spectra to a value (3.1 GeV). A second challenge is the timing alignment among 1,296 channels of 24 calorimeters. Before applying any offline software corrections, all the channels have different digitization start times. Moreover, their relative timing offsets will change after any digitizer frontend initialization. In the following paragraphs, procedures for timing alignment will first be explained. The lost-muon energy-calibration technique that relies on good timing alignment will then be described.

Timing alignment

Timing alignment for calorimeter channels is done with the following steps:

1. Correction for waveform-digitizer timing offsets within a calorimeter and trigger-timing offsets among calorimeters

2. Correction for timing difference in laser-sync pulse propagation time within a calorimeter

3. Correction for timing difference in laser-sync pulse propagation time among calorimeters

4. Absolute timing alignment to the incoming beam

For Step 1, laser-sync pulses that appear almost-simultaneously in each of the calorimeter channels, about 25 μs before the beam arrives, are offset to $t = 0$. All the calorimeter hits are offset by the same amount at the channel-by-channel and fill-by-fill basis. At this step, it is assumed that all the optical fibers delivering laser-sync pulses are of the same length. For Step 2, for each calorimeter, the distribution of time differences $DT(i,j)$ between channels i and channel j sharing the same positron event is used. On average, the time difference histograms $DT(i,j)$ are centered at $DT \approx 0$. Any deviation from zero is attributed to the difference in the fiber length and these deviations are extracted as constant offsets to be applied to the timing alignment. Example histograms for $DT(i,j)$ are shown in Fig. 4.12-1. For Step 3, events where muons escape the storage ring before decaying (the so-called "lost-muons") and hit multiple calorimeters are used, and the calorimeter-to-calorimeter time difference $DT(C_i, C_j)$ of lost-muon events are aligned to ≈ 6.2 ns, which is roughly the time it takes for a muon to travel from one calorimeter to another. A decent energy-scale calibration is needed for this step to select lost-muon events with high signal-over-background ratio. Hence, it is done over several iterations together with the lost-muon energy calibration technique described below. Finally, in Step 4, the time difference between the laser-sync pulse observed in the T0 detector (a scintillator read out by PMTs and SiPM) and the energy-weighted-time of the incoming muon beam $DT(beam, sync)$ is used to correct for the beam-sync pulse jitter in all calorimeter channels.

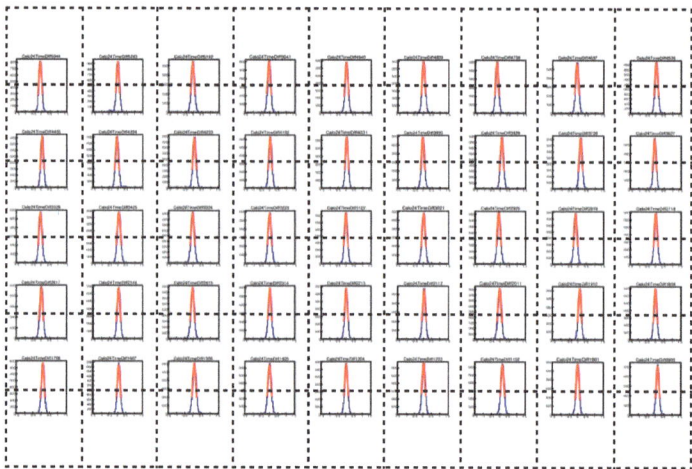

Figure 4.12-1. Shower time-difference spectra between various channels of a calorimeter before Step 2 in the text. Each histogram is the $DT(i,j)$ distribution between channels i and channel j. The peak of each distribution is not at zero initially and any deviation from it is extracted as an offset correction between channel i and j.

Energy scale equalization and calibration

A novel energy equalization and calibration technique was developed for the muon $g-2$ experiment at Fermilab by utilizing "lost-muon" events and Geant4 simulation. By selecting events where muons traverse only a single crystal (evident by no activity in the surrounding crystals) in one calorimeter, and then hit two subsequent calorimeters (also each through a single crystal), the peak of the reconstructed energy for each channel can be used for energy scale equalization. Such a triple-coincidence event strongly suppresses any accidental background of low-energy positron events. At the energy-equalization stage, all the lost-muon peaks are calibrated to 200 MeV. It is challenging to extract the corresponding number from simulation due to the fact that the calorimeter response for high energy muons is very sensitive to tunable parameters that include Cerenkov emission angle, surface roughness, absorption spectra, and crystal light yield. More work is required to get an accurate prediction. The current simulation was optimized for to match past SLAC test-beams for accurate electron response; it predicts a reconstructed energy of about 250 MeV for lost-muons. This prediction is too high when compared with the location of lost-muon peaks determined by an E821-style energy-endpoint calibration.

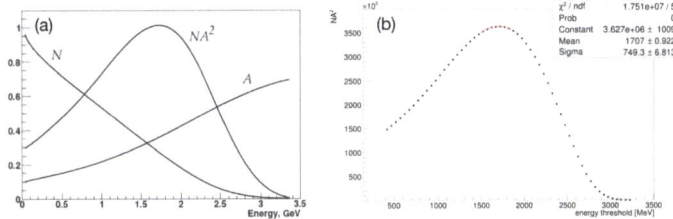

Figure 4.12-2. Absolute scale energy calibration using the NA^2 curve from five-parameter fit, where N is the normalization factor and A the asymmetry of muon decay (see text). (a) NA^2 curve from Geant4 simulation and (b) NA^2 curve from pre-calibrated data.

At the moment, the absolute energy-scale calibration is performed by comparing the NA^2 curve of the standard muon $g-2$ five-parameter fit, $N(t) = Ne^{-t/\gamma\tau_\mu}(1 - A\cos(\omega_a t + \phi))$ to the high-energy positron time spectrum (Sec. 4.11), for the simulation and for the experiment. Here, N is the normalization factor, γ the Lorentz factor, τ_μ the muon rest lifetime, A the asymmetry of muon decay, ω_a the anomalous muon precession frequency and ϕ the phase of the precession. The NA^2 curve is obtained by scanning the energy threshold applied to the positron energy distribution when making the time spectrum. The peak of NA^2 curve from the simulation is at about 1.7 GeV and the data is calibrated to this value. After this absolute-scale calibration step, the lost-muon peaks align at about 170 MeV. Example energy spectra after the calibration procedure are shown in Fig. 4.12-3.

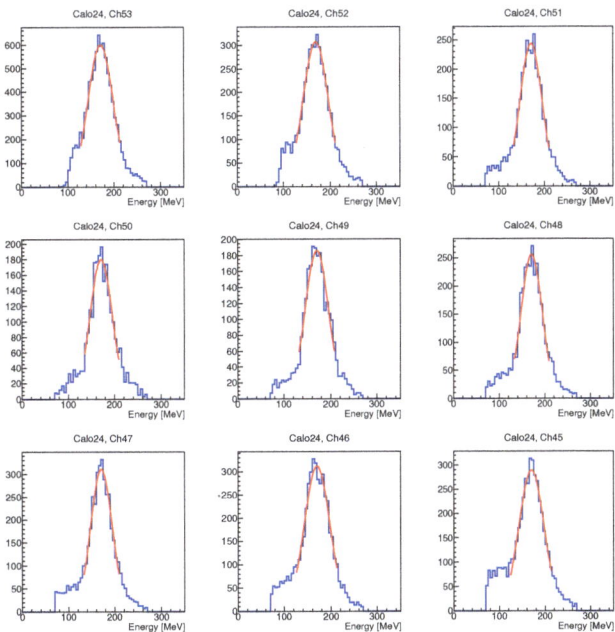

Figure 4.12-3. Energy deposition spectrum of muons which traverse a single crystal as minimum ionizing particles for nine of the calorimeter channels. Initially at different locations, the peaks are aligned at about 170 MeV after applying energy calibration technique using lost-muons and the NA^2 curve.

After performing the timing alignment, energy equalization, and energy-calibration procedures described above, the utility of the reconstructed data is significantly improved. For instance, the previously-smeared-out beam-time profile is now visible thanks to the timing alignment. This enables study of the muon-beam de-bunching dynamics with greater precision. Moreover, the signal-to-background ratio for identifying a lost-muon has improved by a factor of at least 150, compared to the dataset before correction and calibration. This enables the extraction of the number of lost-muons as a function of time with higher precision.

In summary, utilization of laser-sync pulses, lost-muons, and positron events has enabled a novel timing alignment, energy equalization, and energy calibration for calorimeters of the muon $g-2$ experiment. Systematic uncertainties for these procedures are currently under investigation.

4.13 IBMS update

J.F. Amsbaugh, P. Kammel, B.K.H. MacCoy, D.A. Peterson, and T.D. VanWechel

Muon beam injection into the $g-2$ storage ring is a challenging and critical step in muon storage. The entering muon beam must pass through strong fringe fields and narrow inflector apertures (18 mm × 56 mm), which results in a mismatch between transmitted beam phase space and the storage ring's acceptance. The Inflector Beam Monitoring System (IBMS) was designed and constructed at CENPA. It is the primary diagnostic tool to verify the beam optics tune in the muon injection section, and is in use as a continuous monitor of beam properties. Two IBMS modules (IBMS1 and IBMS2) have been built, tested, and installed for operation in the 2018 commissioning run; IBMS3 is planned for future installation.

IBMS1 and IBMS2 consist of two planes of 16 scintillating fibers (SciFis) each, oriented in the horizontal and vertical directions. Each 0.5 mm-diameter, double-cladded fiber (SCSF-78 from Kuraray) is coupled to a silicon photomultiplier (SiPM) (S12571-010P from Hamamatsu). These 1 mm^2-area SiPMs were selected for their large dynamic range (10,000 pixels) and fast recovery time. Detector geometry was optimized for the expected beam profile at each position. The readout electronics are based on low-noise programmable gain amplifiers followed by line drivers which transport the analog signals via micro-coax cables to the receiving CAEN V1742 waveform digitizers. Each 16-SiPM-channel board fits onto one 16-fiber detector plane, with each channel reading out one SciFi. A Sr beta source, along with a special high-gain amplifier board, was used to characterize the detector response to minimum ionizing particles (MIPS). Neutral-density filters placed in front of the SiPMs then set the detector response within the linear SiPM operating range for the expected 700,000 muons per 120 ns beam pulse.

IBMS2: The IBMS2 detector is positioned inside the yoke-hole entrance into the ring magnet, just upstream of the inflector. This tight space is accessed by a linear rail system installed inside the 100-mm-diameter injection pipe. A carriage and winch system carries the IBMS2 detector assembly along the rail to its operating position 2 m downstream of the entrance. The injection pipe imposes tight space restrictions in the detector plane and limitations on thermal management. Fiber pitch is 3.25 mm in both the X and Y planes, for a 52 mm × 52 mm active area.

Figure 4.13-1. IBMS2 consists of 16 × 16 SciFis on a 3.25 mm pitch. The SciFis are read out by two 16-channel amplifier and SiPM boards. IBMS2 is located inside the injection pipe into the storage ring, just upstream of the inflector aperture into which the beam is steered. The view into the injection pipe (*right*) shows IBMS2 at its operating position 2 m downstream.

IBMS2 was installed in June 2017 for operation in the initial commissioning run, and successfully provided real-time beam profiles during the run. Upgrades during the summer included a newly-built fiber frame, featuring straighter fibers than the first version, and a light-injection system for stability monitoring. The light injection system couples pulsed surface-mount LEDs to a notched optical fiber; the notches direct LED light into each SciFi, allowing real-time monitoring for each trigger. After these upgrades, IBMS2 was re-installed in fall 2017, along with IBMS1, for operation in the 2018 commissioning run.

IBMS1: The IBMS1 detector is the upstream-most of the IBMS detectors, positioned just outside the yoke hole (just downstream of the T0 detector). Due to its position, space is constrained along the beamline, but is relatively unconstrained in the detector plane. IBMS1 is mounted on an X-Y translation stage, which allows remote positioning in the detector plane for fine adjustments and for displacement by known amounts for in-beam detector calibration. Fiber pitches are 5.5 mm (X plane) and 2.7 mm (Y plane), for an 88 mm × 43.2 mm active area.

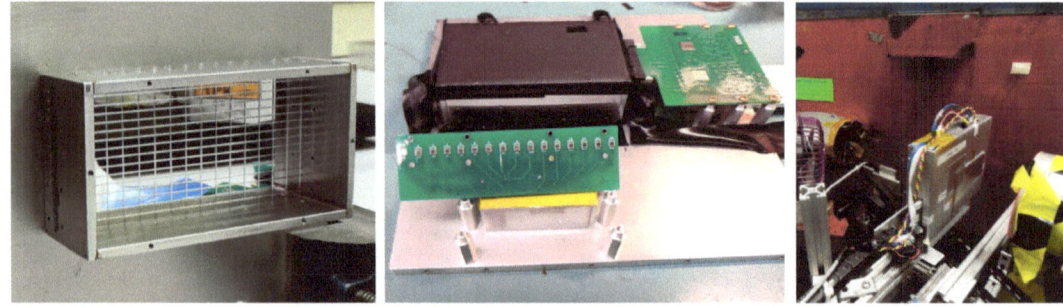

Figure 4.13-2. IBMS1 consists of 16 × 16 SciFis on 5.5 mm (X) and 2.7 mm (Y) pitches. The SciFis are read out by two 16-channel amplifier and SiPM boards, each of which is a modular right-angle assembly. IBMS1 is located approximately 2.4 m upstream of IBMS2, just outside the yoke-hole entrance to the ring, before the beam is strongly affected by fringe fields. The IBMS1 assembly is shown at its operating position just outside the yoke-hole (*right*).

Readout electronics are functionally the same as the IBMS2 electronics, but feature independent SiPM and amplifier boards joined with a right-angle connector. This modular design allows easy swapping of the amplifier boards for characterization and repair. A light injection system, similar to the IBMS2 system, allows stability monitoring of the IBMS1 response. IBMS1 was installed in fall 2017, and has been successfully operating in the 2018 commissioning run. Initial beam profiles for IBMS1 and IBMS2 are shown in Fig. 4.13-3.

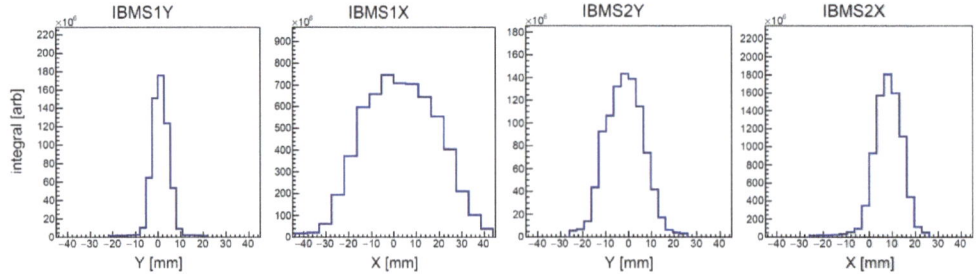

Figure 4.13-3. Spatial beam profiles for IBMS1 (*left*) and IBMS2 (*right*) during the commissioning run; Y profiles show the vertical spatial distribution of the beam, and X profiles show its horizontal spatial distribution. Average spatial beam profiles are given by time integrals of the 120-ns-long beam pulse for each fiber in the plane. The profiles show that the spatial distribution of the beam evolves along the injection path, from short and wide at IBMS1 (with $2\sigma_y = 7.4$ mm, $2\sigma_x = 33.4$ mm) to somewhat tall and narrow at IBMS2 (with $2\sigma_y = 15.6$ mm, $2\sigma_x = 11.0$ mm).

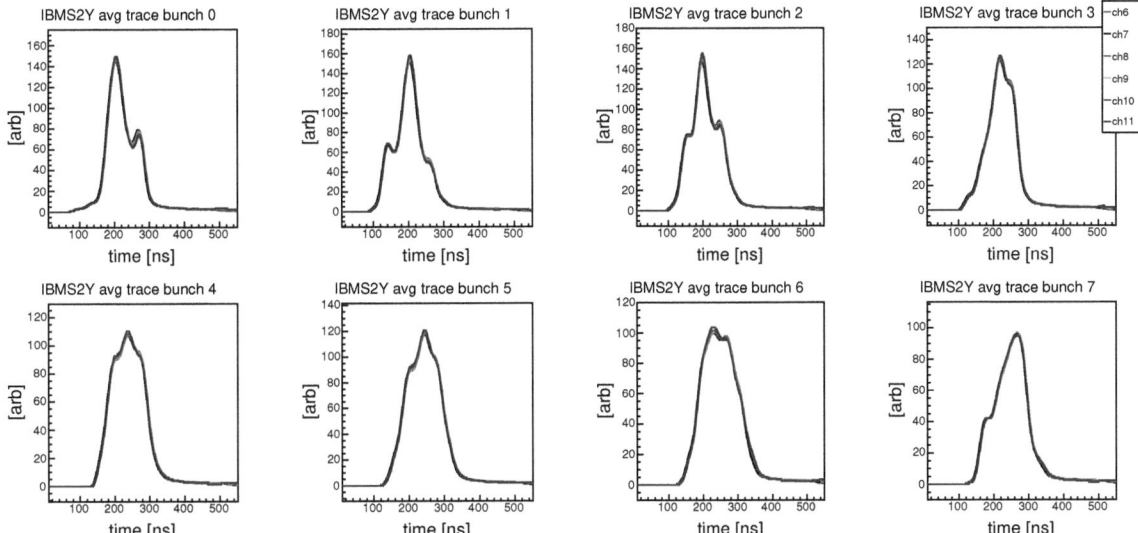

Figure 4.13-4. Average temporal beam profiles (arbitrary units) from IBMS2Y, for the 8 beam bunches in an accelerator cycle. Average waveforms for channels 6-11, corresponding to the 6 hottest fibers near the beam centroid, are normalized by time integral and overlayed. The waveforms show similar temporal shape for each channel (bunch by bunch). The similarity of channels which see different light intensities indicates linear detector response, while the similarity of fibers at different positions shows time stability of the spatial beam profile over each 120-ns-long beam pulse.

DAQ and analysis software

DAQ and analysis software has been developed with help from CENPA group members Smith, Fienberg, and Khaw. The MIDAS-based DAQ software allows the IBMS to run in the experiment's main DAQ system. Analysis software for beam profile characterization includes standalone C++ analysis at CENPA, and analysis in the $g-2$ offline code based in the ART framework.

4.14 Monitoring the $g-2$ storage ring magnetic field with fixed probes

M. Fertl, A. García, R. Osofsky, and H. E. Swanson

The storage-ring magnetic field is measured using Larmor precession from pulsed proton NMR. Individual probes contain proton samples and coils to excite and sense the resulting free induction decays (FIDs). Measurement of the field in the muon-storage volume uses a movable trolley containing an array of NMR probes which follows muon trajectories. As the trolley blocks the beam, the field is continuously monitored for field drift by so called Fixed Probes which occupy 378 fixed locations above and below the muon-storage volume. These sample the ring field every 5 degrees in azimuth and provide continuous measurements as proxies for the trolley while muons are stored in the ring. The probes, and their associated readout electronics, were designed and constructed at CENPA[1]. They have been in operation

---

[1] CENPA Annual Report, University of Washington (2017) p. 108.

for over a year. Early in system commissioning, a maximum rating was exceeded causing premature failure of some components. Most of these have been replaced and system uptime is near 100 percent.

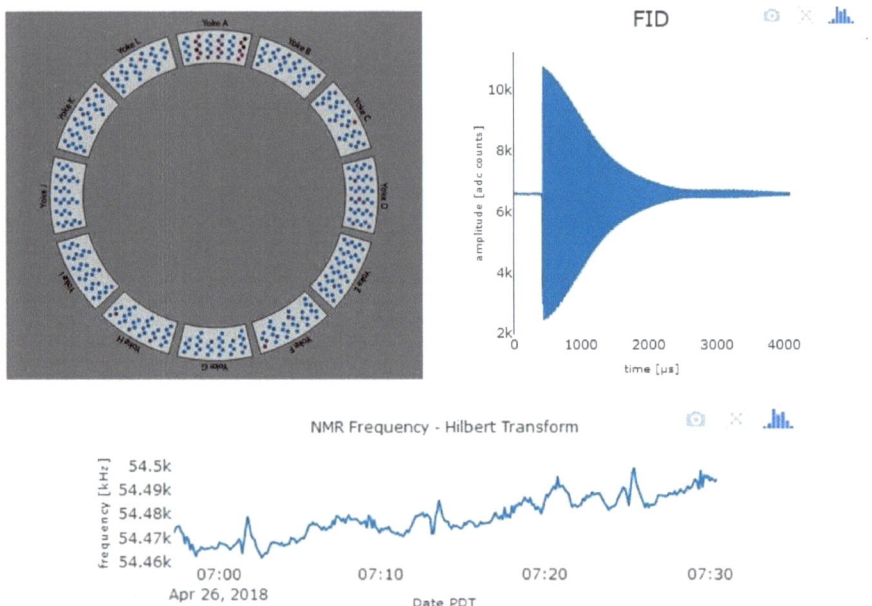

Figure 4.14-1. Screen-shots from the Fixed Probe on-line monitor. *Upper-left* indicates the quality of FID waveforms for each of the probes (blue - good, red - moderate, black - poor), *upper-right* is a typical good quality FID waveform from a selected probe and the *lower* plot shows the history of extracted frequencies from that probe.

The FID waveforms from all probes are digitized and their Larmor precession frequencies extracted every 2 seconds. A subgroup of probes is used for feedback to the magnet power supply which holds their average value constant to a few parts per billion (ppb). On-line displays keep the experiment-monitoring "shifters" aware of proper operation. Fig. 4.14-1 shows screen shots from one of these displays. Upper-left is an overview graphic showing each probe's measurement efficacy. Strong gradients in a probe's vicinity shorten the FID's decay time and reduce the precision with which frequencies can be extracted. The colors blue, red, and black indicate the existence of a FID waveform and whether its length is sufficient for extraction algorithms. All but about 20 probes achieve a precision of 70 ppb or better. The other two plots show the FID wave form from a selected probe and its recent history of extracted frequencies.

Due to their location and proximity to iron shims, fixed probes experience different fields (of order a few ppm) from those seen by muons at the same azimuthal location. To function as a proxy for trolley-measured fields, they must at least respond to field drifts in the same manner. This is demonstrated by comparing the difference in azimuthal field maps from two trolley measurements with corresponding differences in fixed probe values. When the trolley passes by there is a significant perturbation in the field measured by the fixed probe. Measurements taken before and after the trolley passes are interpolated to determine a value for the unperturbed field. The upper panel in Fig. 4.14-2 shows the difference in storage-region field measured at two different times. The lower panel gives the equivalent difference measured outside the storage volume by the fixed probes. A comparison of these plots show that general features and overall amplitudes behave the same in response to drifts in the field. The fixed probes are, on average, seen to be a good proxy for the dipole field as measured by the trolley.

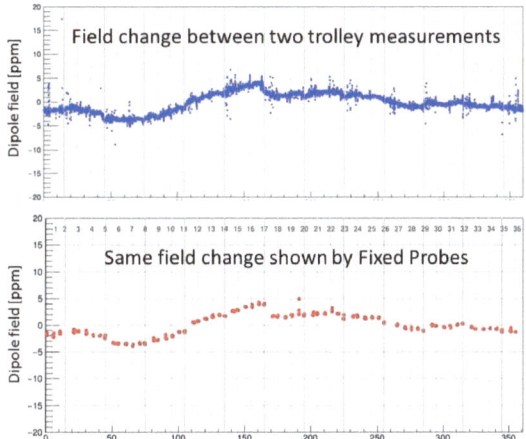

Figure 4.14-2. Top: The difference between two trolley field measurements as a function of azimuth. Bottom: The corresponding difference between fixed-probe measurements made at the times of the trolley measurements.

With its array of 17 probes the trolley also measures higher order field multipoles. At each of the 72 azimuthal fixed-probe locations there are 4 to 6 probes. This spatial coverage provides some sensitivity to the field's quadrupole and sextapole moments. We are currently investigating how well the fixed probes can predict changes in field multipoles measured by the trolley.

4.15 Active shimming of the $g-2$ storage ring magnetic field using surface coils

M. Fertl, A. García, R. Osofsky, and H. E. Swanson

The targeted systematic error on the magnetic field in the muon $g-2$ experiment is 70 parts per billion (ppb). To achieve this goal, the magnetic field must be as homogeneous as possible, ideally less than ± 1 ppm across the storage region. The process of smoothing the field, known as shimming, is split into 2 distinct processes: passive shimming and active shimming. Passive shimming, which took place from September 2015 - August 2016, involved the adjustment/addition of over 10,000 pieces of ferrous and nonferrous material in the muon storage region. Active shimming, which began in Spring 2017, is continually ongoing.

The magnetic field in the muon storage region is characterized in terms of 2D multipoles, defined by

$$B(r,\theta) = B_0 + \sum_{n=0}^{4} \left(\frac{r}{r_0}\right)^n [a_n \cos(n\theta) + b_n \sin(n\theta)], \qquad (1)$$

where r and θ are the polar coordinates of the storage region cross section. In April 2017, after passive shimming had been completed and all vacuum chambers had been installed in the muon storage region, the magnetic field varied by ±6 parts per million (ppm) across the storage region, with a -3.8ppm normal octupole moment, which by convention has a six-fold spatial symmetry, as the dominant multipole (see left of Fig. 4.15-1).

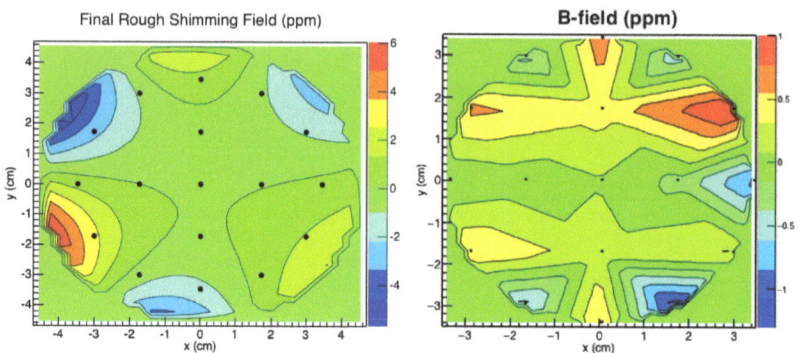

Figure 4.15-1. Surface coils are used to minimize the variations across the azimuthally averaged storage region magnetic field. The final rough-shimming magnetic field, from April 2017, is pictured on the left with variations of ±6ppm across the storage region. The April 2018 field, with variations of only ±1ppm, is shown on the right.

Active shimming has two components: power-supply feedback and surface coils. Power-supply feedback refers to using a current source to adjust the main magnet current so as to keep the dipole field in the storage region (B_0 in Eq. 1) constant to within ±100ppb. Surface coils are two sets of 100 concentric current-carrying coils located above and below the vacuum chambers which are used to minimize the azimuthally averaged multipoles. The electrical-current range of all 200 coils is ±2.5A.

By varying the currents in the coils as a function of radial position, different multipoles can be targeted, with the configurations shown in the table below where a is a constant and x is radial position. While each configuration targets 1-multipole moment, each affects all multipole moments. The magnetic-field measuring 'trolley was used to make measurements of the effects of each configuration on all field multipoles.

Multipole	Normal, top	Normal, bottom	Skew, top	Skew, bottom
Dipole	−	−	1	1
Quadrupole	a	a	x	$-x$
Sextupole	ax	ax	$x^2 - a^2$	$-x^2 + a^2$
Octupole	$3ax^2 - a^3$	$3ax^2 - a^3$	$x^3 - 3a^2x$	$-x^3 + 3a^2x$
Decupole	$ax^3 - a^3x$	$ax^3 - a^3x$	$x^4 + a^4 - 6a^2x^2$	$-x^4 - a^4 + 6a^2x^2$

In addition to affecting the main field multipoles, each of the surface coil configurations also affects the radial field in the storage region. Because NMR probes measure field magnitude, not field direction, the trolley could not be used to measure the effect of each configuration on the radial field. Instead, the effect was calculated using the Biot-Savart law and the method of magnetic images, where the coils were assumed to be infinitely long and straight and the pole pieces were assumed to have very large permeability. Calculated multipole effects for each configuration matched the trolley measurements to within a couple ppm for very large imposed gradients.

Once the field effects of each multipole configuration were well understood, the currents in all 200 coils were optimized using a least-squares optimization method. The final azimuthally-averaged field in the storage region varies by ±1ppm across the storage region, with all multipoles smaller than 0.25ppm, and is shown on the right of Fig. 4.15-1.

5 Dark matter searches

ADMX (Axion Dark Matter eXperiment)[1]

5.1 Overview of ADMX

R. Cervantes, N. Du, N. Force, R. Khatiwada, S. Kimes, E. Lentz*, R. S. Ottens†, L. J. Rosenberg, G. Rybka, and D. I. Will

The Axion Dark Matter eXperiment (ADMX) is a search for galactic dark matter axions whose detection apparatus is an implementation of Pierre Sikivie's axion haloscope[2]. The ADMX collaboration has operated a series of axion searches since the mid-1990s. The experiment has been located at CENPA since 2010. Following a data run in 2013-2014, ADMX has been undergoing a series of upgrades, most notably the addition of a dilution refrigerator, to enter phase "Gen 2." At the time of writing, March 2018, ADMX Gen 2 has completed and published the results of its first science run where it achieved an unprecedented level of sensitivity to dark matter axions. ADMX Gen 2 has begun a second science run using new quantum amplifiers to probe a different range of axions.

The axion first emerged in the late 1970s as a consequence of a solution to the "strong CP problem." The strong CP problem can be concisely stated as: An exceedingly fine, arguably unnatural, tuning of the Standard Model is required to account for the conservation of the discrete symmetries P (parity) and CP (charge conjugation times parity) within quantum chromodynamics (QCD). The amount of CP violation in QCD is encoded in a phase θ which appears in the QCD Lagrangian. When θ differs from zero, QCD violates P and CP. Since the strong interactions appear to be P- and CP-symmetric in the laboratory, θ must be very small, namely, the upper limit on the neutron electric-dipole moment requires $|\theta| < 10^{-10}$. However, in the Standard Model, P- and CP-violation by the weak interactions feeds into the strong interactions so that the expected value of θ is of order unity. Peccei and Quinn proposed a solution to this problem in which the Standard Model is modified so that θ becomes a dynamical field and relaxes to zero, thus conserving CP in a natural way[3]. The theory's underlying broken continuous symmetry dictates existence of a particle: the axion. The axion is the quantum of oscillation of the θ field and has zero spin, zero electric charge, and negative intrinsic parity. So, like the neutral pion, the axion can decay into two photons. Soon after the initial proposal of the axion, it was realized within a certain range of possible masses the axion is ideal dark matter candidate. Experiments and astronomical observations have since constrained the axion's mass to within a range in which the axion contributes a significant portion of dark matter; this mass range corresponds to microwave photons.

---

*Presently a Postdoc at University of Göttingen.
†Presently a Research Scientist at Raytheon Corp.

[1] ADMX is supported by the DOE Office of High-Energy Physics and makes use of CENPA resources by recharge to the service center.
[2] P. Sikivie, Phys. Rev. Lett. **51**, 1415 (1983).
[3] R. D. Peccei and H. R. Quinn Phys. Rev. Lett. **38**, 1440 (1977).

ADMX operates by stimulating the conversion of local dark matter axions into photons via Primakoff conversion and detecting the resulting photons. The experiment consists of a large tunable microwave cavity (Fig. 5.1-1), immersed in a static magnetic field. The resonance of the cavity further enhances the conversion, and the resonance is tuned to search for axions of varying mass. An RF receiver records the power spectrum of the cavity for each tuning configuration as ADMX scans in frequency. However, despite the prodigious predicted local density of dark matter axions (approximately 10^{14}/cc), the expected electromagnetic signal is extraordinarily weak, around 10^{-22} W in the ADMX apparatus. Our present ADMX collaboration (University of Washington, Lawrence Livermore National Laboratory, University of Florida, University of Sheffield, National Radio Astronomy Observatory, University of California Berkeley, Pacific Northwest National Laboratory, Los Alamos National Laboratory, Fermi National Accelerator Laboratory, and Washington University) has constructed and is operating the first, and presently only, experiment sensitive to plausible dark matter axions.

ADMX is essentially an extraordinarily low-noise radio receiver with a tunable RF cavity forming a tuned resonance circuit. A short electric-field probe couples power from the cavity into a cryogenic amplifier which is cooled to near the cavity temperature.

The dominant background in ADMX comes from thermal noise; the system noise temperature is the noise temperature of the amplifier plus the cavity physical temperature. The motivation for lowering the system noise temperature is clear: (i) for a given axion-photon coupling, the frequency scan rate decreases as the square of the system temperature and (ii) for a given scan rate, the power sensitivity increases as the system temperature drops. ADMX reduces its thermal noise by using low noise first stage amplifiers and cooling the cavity. ADMX collaboration members developed superconducting quantum interference device (SQUID) amplifiers in the $100 - 1000$ MHz range specifically for ADMX; this development allowed more than an order-of-magnitude reduction in system noise temperature.

The previous version of ADMX, ADMX Phase 1, was located at Lawrence Livermore National Laboratory and operated from 2002-2009. The Phase 1 apparatus was fitted with SQUID amplifiers and operated at approximately 2 K physical temperature via a pumped ^{4}He system. ADMX Phase 1 completed a scan of the $1.9 - 3.5$ μeV axion-mass range and the results for conservative estimates of dark matter density were published[1]. After further analysis the ADMX collaboration published a search for axions assuming non-virialized dark matter distribution within our galaxy[2].

Starting in 2010, ADMX was entirely redesigned and rebuilt at CENPA. The 2013-2014 data runs of ADMX (hereforth referred to as "Phase 2a") were the first data collected by ADMX at CENPA. Throughout the 2014 data runs ADMX employed a pumped ^{4}He system achieving physical temperatures of about 1.2 K and SQUID amplification. The success of Phase 2a demonstrated full functionality of the new ADMX design and success of the closed loop helium liquefaction system in maintaining ADMX operations through order-months extended data runs. Additionally, Phase 2a was the first time an axion haloscope has taken

---
[1]S. Asztalos *et al.*, Phys. Rev. Lett. **104**, 041301 (2010).
[2]J. Hoskins *et al.*, Phys. Rev. D **84**, 121302 (2011).

Figure 5.1-1. *Left:* Removal of the ADMX insert after a successful commissioning run *Right:* ADMX insert in the cleanroom.

data through two independent receiver chains simultaneously and the first time a nonfundamental cavity mode achieved sensitivity to plausible QCD axions. New axion exclusion limits were set in two frequency regimes: 600 – 720 MHz (corresponding to the "TM010" mode) and 1050 – 1400 MHz ("TM020" mode), with benchmark QCD axion sensitivity in select frequency ranges.

ADMX was selected as a flagship DOE "Gen 2" dark matter experiment in fall 2014 and began a period of design, build, and upgrade. Among the major upgrades are: the installation of a dilution refrigerator and supporting plumbing and infrastructure, a redesigned RF system including Josephson Parametric Amplifier and SQUID amplifiers, and substantial redesign of the thermal linkages and connections in the experiment to achieve lowest possible operating temperature.

ADMX Gen 2 concluded its first science run in June 2017, searching for axions in a 645-680 MHz range with SQUID amplifiers. In that range, axion-like signals were not observed so instead, 90% upper confidence limits were placed on the axion-photon coupling over the search range. These limits excluded DFSZ axions between 645-676MHz, representing the first time an axion haloscope has excluded axion couplings with DFSZ sensitivity[1]. With this sensitivity, ADMX has become the only operating experiment to probe the DFSZ grand unified theory coupling for the axion.

Currently, ADMX is in the middle of a second data taking run using a combination of a Josephson parametric amplifier and SQUID amplifier to search a much larger range from

---

[1] N. Du *et al.*, Phys. Rev. Lett. **120**, 151301 (2018).

680-800 MHz. The magnet is being operated at a higher field of 7.6 T compared to 6.8 T in the previous run. In addition, changes to the plumbing of the dilution refrigerator and the thermal linkages of the experiment allow it run with the mixing chamber at 90 mK, compared to 150 mK in the previous run. Together, these improvements will allow the experiment to probe for axions at a faster rate. The collaboration is in a position to discover the axion at any moment.

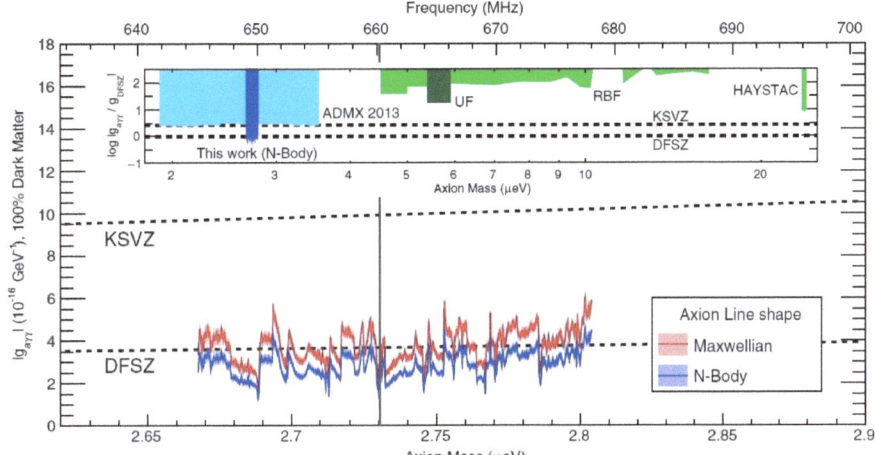

Figure 5.1-2. Limits on axion-photon couplings set by the ADMX Gen 2 following its first science run in 2017. The limits were made assuming that axions composed 100% of dark matter in the universe. Due to external radio interference from 660.16 to 660.27 MHz, we did not set limits in that mass range, indicated by the gray bar. These limits represent the first time an axion haloscope has been capable of probing DFSZ grand unified theory coupling for the axion[1].

5.2 Higher-frequency axion searches with Orpheus

R. Cervantes, S. Kimes, R. S. Ottens[†], and G. Rybka

Axions are hypothetical particles that, if they exist, would solve both the strong CP problem and the dark matter problem. Axions in our local dark matter halo could be detected using an apparatus consisting of a resonant microwave cavity threaded by a strong magnetic field. The ADMX experiment has recently used this technique to search for axions in the few $\mu eV/c^2$ mass range. However, the ADMX search technique becomes increasingly challenging with increasing axion mass. This is because higher masses require smaller-diameter cavities, and a smaller cavity volume reduces the signal strength. Thus, there is interest in developing more sophisticated resonators to overcome this problem.

The ADMX Orpheus prototype experiment aims to search for axion-like particles with masses approaching 100 $\mu eV/c^2$. The Orpheus experiment consists of a dielectric-loaded Fabry-Perot resonator. The Fabry-Perot cavity operates at a high-order mode in order to

---

[†]Presently a Research Scientist at Raytheon Corp.

increase the mass range while keeping the detection volume high. The open design reduces dissipation in the cavity walls, thus increasing Q factor. Alumina lenses are placed every fourth half-wavelength in order to suppress the electric field that is anti-aligned with the magnetic field, increasing the cavity's coupling to the axion. Fig. 5.2-1 shows the current design of Orpheus[1,2], and Fig. 5.2-2shows the planned parameter space that the Orpheus prototype will explore.

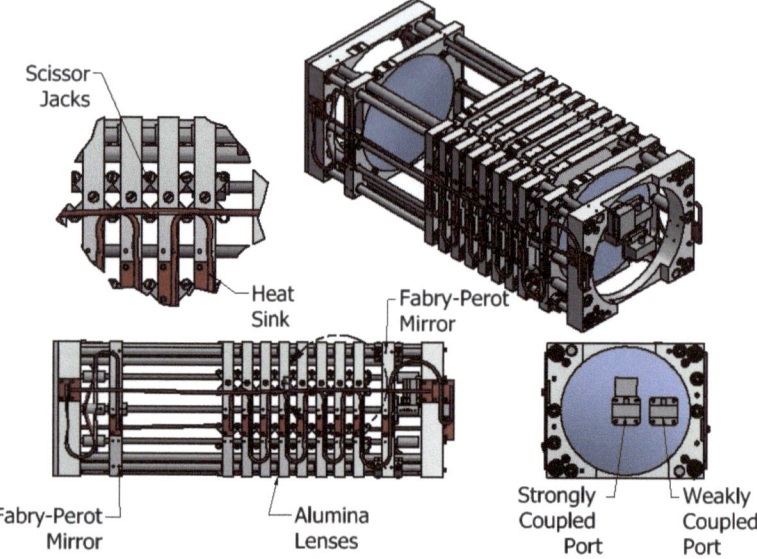

Figure 5.2-1. The mechanical design for Orpheus. It is a Fabry-Perot cavity with alumina lenses evenly spaced throughout. Orpheus will sit in a 1 T uniform dipole magnet. This design is a work in progress and is not the final design.

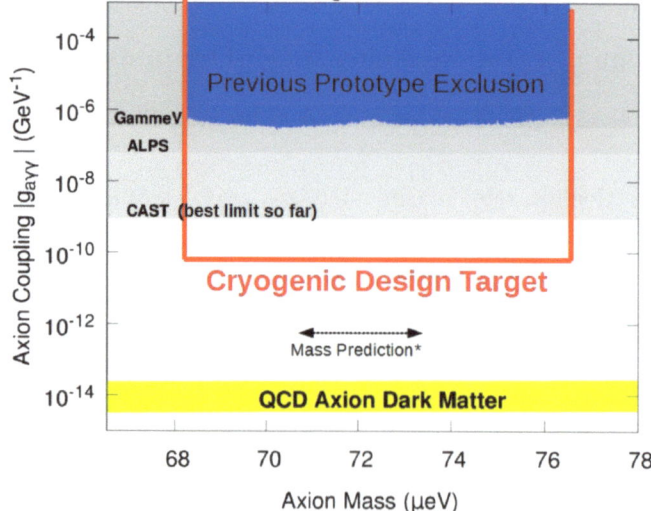

Figure 5.2-2. The scientific reach of the Orpheus experiment.

---

[1] Phys. Rev. D **91**, 011701 (2015).
[2] L. Visinelli and P. Gondolo, Phys. Rev. Lett. **113**, 011802 (2014).

We have simulated a miniature Orpheus experiment that is five half-wavelengths long. The mirrors are 1 inch in diameter, with WR-51 waveguides attached. A 3 mm feedthrough hole is drilled into the mirrors so that the waveguide can couple to the cavity. The cavity is in a plano-spherical configuration, meaning that one mirror is flat while the other is a spherical mirror. We have chosen the radius of curvature of this mirror to be eight times the optical length, so that the gaussian beam will have a large waist throughout the cavity.

The forward transmission coefficient S_{21} is simulated with different Fabry-Perot cavity configurations: without dielectrics (empty space), with flat dielectrics, and with dielectrics possessing the same curvature as one of the Fabry-Perot reflectors. From the S_{21} simulations, one can determine the Q of the resonator (Fig. 5.2-3). We find that Q is better with when the cavity is loaded with dielectrics, and even better with when the dielectrics are curved (Table 5.2-1).

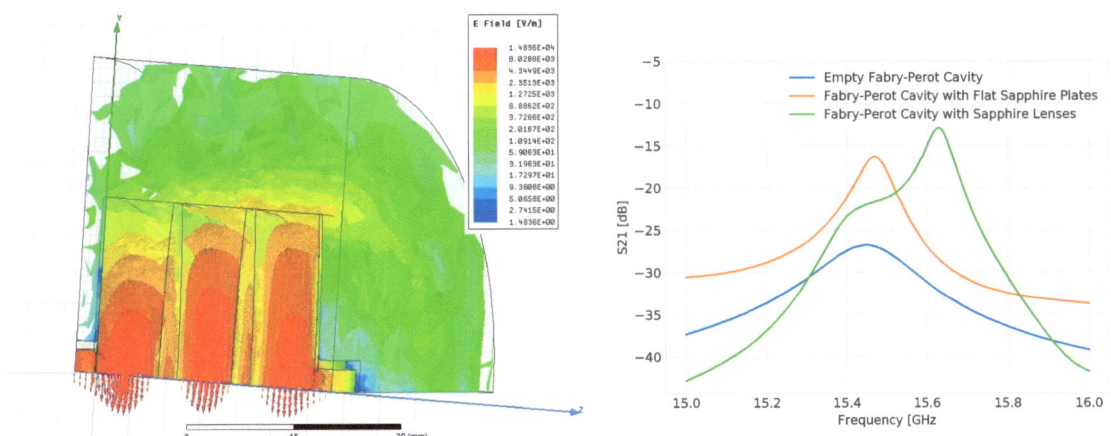

Figure 5.2-3. Left: A simulated miniature Orpheus. A gaussian beam resonantes inside of the Fabry-Perot cavity. The dielectric lenses both focus the gaussian beam and suppress the field where it is anti-aligned with the dipole magnetic field. Right: From the simulation, the forward transmission coefficient S21 is calculated. The Q-factor is then determined from S21.

Orpheus Configuration	Q Factor
empty cavity	65.9
flat dielectrics	175.6
curved dielectrics	223

Table 5.2-1. The Q factor of a simulated Orpheus for various configurations. The trend suggests that loading the Fabry-Perot cavity with dialectrics increases Q, and curving the dielectrics increases the Q even more.

We are now working towards simulating a larger geometry and optimizing the geometric parameters to maximize the Q factor.

DAMIC

5.3 DAMIC: dark matter in CCDs

A. Burgess, A. E. Chavarria, P. Mitra, A. Piers, and X. Tang

The DAMIC experiment at SNOLAB[1] employs the bulk silicon of scientific-grade charge-coupled devices (CCDs) as a target for interactions of particle dark matter from the galactic halo. By virtue of the low readout noise of the CCDs, DAMIC is particularly sensitive to the small ionization signals (only a few ionized carriers or e^-) from recoiling nuclei[2] or electrons[3] following the scattering of dark matter with masses in the range 1 MeV–10 GeV.

The large-area mm-thick active volume of the CCDs is sensitive to ionizing radiation: both the dark matter signal and backgrounds from natural radioactivity. To mitigate backgrounds, the devices are deployed in a radio-pure tower deep underground, heavily shielded from environmental radiation.

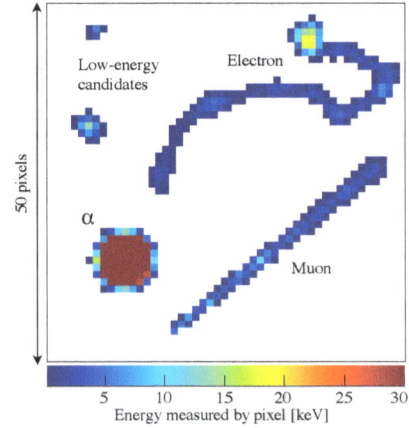

Figure 5.3-1. *Left:* A 16-Mpixel DAMIC CCD in its low radioactivity copper holder. The device area is $6\times 6\,\text{cm}^2$. The CCD is wire bonded to a flex cable that carries the signals that drive the device. *Center:* The DAMIC copper modules slide into slots of a copper box that is cooled to 130 K in a vacuum chamber. *Right:* A segment of a CCD image acquired in the surface laboratory. The pixel size is $15\times 15\,\mu\text{m}^2$. Different tracks can be observed from different ionizing particles. The "low-energy candidates" look like those that could arise from the interactions of dark matter particles.

The CCD images contain a two-dimensional projection on the x-y plane of all the ionization that ocurrs in the active volume of the device. The characteristics of the pixel clusters readily identify the nature of the ionizing particles, which provides important information on their origin. For low energy events (i.e., dark matter candidate events), the track length of the ionizing particle is much smaller than the pixel size and the shape of the cluster is

---

[1] A. Aguilar-Arevalo *et al.*, Phys. Rev. D **94**, 082006 (2016).
[2] J. D. Lewin, and P. F. Smith, Astropart. Phys. **6**, 87 (1996).
[3] R. Essig *et al.*, J. High Energy Phys. **05**, 046 (2016).

determined by how much the charge diffuses as it is drifted along the z direction to the pixel array. Thus, from the measurement of the charge diffusion, it is possible to obtain the z coordinate and, hence, reconstruct in 3-D the location of a particle interaction in the active volume.

DAMIC at SNOLAB

Since February 2017, we have been operating a seven-CCD array (target mass of 40 g) in the J-Drift of the SNOLAB underground laboratory. The CCDs are operating with optimal performance of $1.6\,e^-$ RMS pixel noise and a radioactive background at low energies of 5 events/keV/kg/day. As of March 2018[1], we have acquired 7.6 kg-d of data optimized for maximum spatial resolution and 4.6 kg-d of data optimized for sensitivity to small ionization signals, with a threshold of $15\,e^-$. We are currently analyzing these data to search for dark matter particles (e.g., by WIMP-nucleus elastic scattering and hidden photon absorption), millicharged cosmic particles, and to perform background studies that will inform the design of next-generation silicon dark matter detectors.

DAMIC-1K

The next step in the DAMIC program is a 50-CCD detector array of 1 kg target mass. It capitalizes on the DAMIC experience at SNOLAB while taking a giant leap forward in sensitivity by radically innovating the detector technology. Its 36 Mpixel CCDs will be the most massive ever built, 20 g each. The implementation of a novel "skipper" readout[2] will result in the high-resolution detection of a single charge carrier (e^-). Together with the remarkably low leakage current of $<10^{-21}\,\text{A/cm}^2$ — a combination unmatched by any other dark matter experiment — DAMIC-1K will feature a threshold of 2 or $3\,e^-$.

DAMIC-1K will search for low-mass dark matter in a broad range of masses from 1 eV to 10 GeV. In addition to progress in the search for GeV-scale WIMP dark matter and hidden-photon dark matter, DAMIC-1K will break new ground in the search for dark matter with masses 1 MeV to 1 GeV by improving by orders of magnitude the sensitivity to the ionization signals from the scattering of dark matter particles with valence electrons. The science reach of DAMIC-1K was compiled, together with the prospects for other direct detection experiments, in Section IV of the "Cosmic Visions" community report[3].

---

[1] M. Settimo *et al.*, arXiv:1805.10001 (2018).
[2] J. Tiffenberg *et al.*, Phys. Rev. Lett. **119**, 131802 (2017).
[3] M. Battaglieri *et al.*, arXiv:1707.04591 (2017).

5.4 CCD packaging and testing laboratory

A. Burgess, <u>A. E. Chavarria</u>, P. Mitra, A. Piers, and X. Tang

The first development run of DAMIC-1K CCDs was designed by S. Holland at Berkeley Lab and will be fabricated by DALSA during the summer of 2018. This run includes the first prototype large-area skipper CCDs for DAMIC-1K, which will be delivered directly to the University of Washington. At CENPA, we have set up a laboratory to package, test and characterize these CCDs.

The laboratory is housed in the repurposed clean room from MAJORANA, upgraded with all the relevant precautions to mitigate electrostatic discharge that could damage the devices during handling. It consists of a packaging station where the CCD bare die and the flex cable that brings the signals to and from the CCD are glued in place, with the connection pads on the die and the flex cable aligned for wire bonding. A refurbished automatic wedge bonder K&S 1470 with $10\,\text{cm} \times 10\,\text{cm}$ table travel was procured and is used to make all the wire connections. The entire packaging procedure was validated with CCDs from a previous DAMIC fabrication run. Photographs of the clean laboratory are shown in Fig. 5.4-1, while a CCD on the wire bonding stage and the resulting wire bonds are depicted in Fig. 5.4-2.

 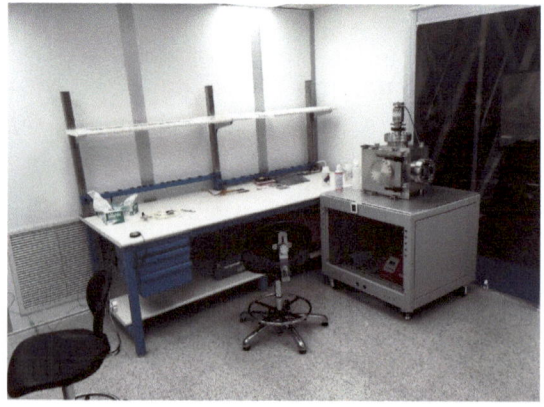

Figure 5.4-1. *Left:* K&S 1470 automatic wire bonder and inspection microscope. *Right:* CCD test chamber under commissioning and CCD mock-ups.

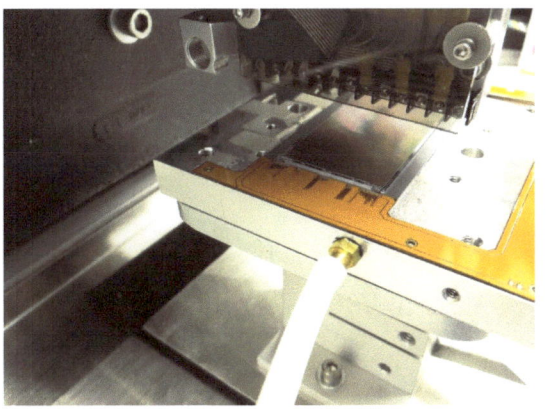 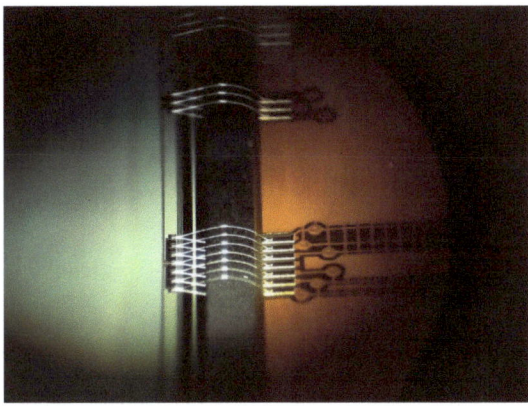

Figure 5.4-2. *Left:* The first 8-Mpixel DAMIC CCD to be wire bonded at the University of Washington. This step completes the CCD package. *Right:* Some of the resulting wire bonds seen through a microscope. Each CCD has 78 pads that must be connected.

A CCD test station that includes a vacuum chamber, a cryocooler and a CCD controller is currently under commissioning. The procurement, fabrication and testing of all components is nearing completion, with the full integration of the system to take place before the summer of 2018. The goal is to use this system to confirm the low-noise performance of the upcoming DAMIC-1K devices, and to directly measure their response to various radiation sources that mimic the signals and backgrounds expected in the dark matter search.

6 Education

6.1 Use of CENPA facilities in education and coursework at UW

J. R. Pedersen

CENPA continues to maintain a prominent role in a broad range of practical, hands-on training and education for both undergraduate and graduate students at the University of Washington. We provide a unique opportunity for students to participate in ongoing research and engineering, resulting in contributions to both local and off-site experiment collaborations.

Once trained, students utilize our experimental laboratories, electronics shop, student machine shop, and chemistry laboratories for their education and collaborative contributions. Complimented by mandatory safety training, students acquire a broad range of practical laboratory skills and knowledge, giving them an significant edge in both academia and industry (Sec. 6.2).

We've also sustained a continued presence in the UW curriculum since 2011, and currently offer an accelerator-based laboratory course in nuclear physics (Phys 575/576), titled "Nuclear Physics: Sources, Detectors, and Safety" (Sec. 6.3).

6.2 Student training

J. F. Amsbaugh, G. T. Holman, D. R. Hyde, J. R. Pedersen, D. A. Peterson, E. B. Smith, T. D. Van Wechel, and D. I. Will

CENPA provides students with training on equipment and best practices in our shops, chemistry laboratories, and experimental laboratories. Prior to that, students receive specific training in laboratory safety, which includes required UW provided in-person and online classes.

Figure 6.2-1. (*Top Left*): Undergraduate student Henry Gorrell working on a CAD design for an ion source filament holder (*Top Right*). (*Bottom Left*): Undergraduate student Grant Leum testing a mu-metal zeroing chamber he designed and built at CENPA. (*Bottom Right*): Undergraduate student Santos Zaid analyzing temperature stability data for a High Purity Germanium gamma spectroscopy system he constructed.

CENPA faculty and staff also provide hands-on student instruction (Fig. 6.2-1) in an array of laboratory technical skills and safety on topics ranging from CAD design, electrical power, compressed gases, cryogenic fluids (liquefaction, dispensing, delivery, and transport), lifting and rigging, vacuum technology, high voltages, high currents, high magnetic fields, nuclear instrumentation, radiation detection and measurement, and radioactive sources. Completion of available UW-provided online safety training is also required.

Our electronics shop is available for use by any trained students (Sec. 7.4) wanting to learn electronic design and assembly. Highly experienced staff members are available on-site to provide instruction in soldering, wiring, and the use of basic electrical and electronic components. Assistance in the use of electronic design automation software for producing prototype printed circuit boards (PCBs) is also available upon request. Additionally, training in the use of our UV ProtoLaser (purchased with UW Student Technology Fee funds) is given to any UW student wanting to fabricate prototye PCBs for research projects (Fig. 6.2-2).

The CENPA student shop facilitates and encourages training for faculty, staff, and students in machine tool operation and safety. A retired staff instrument maker provides instructional machining classes on a weekly basis. Faculty, staff, and students learn how to safely operate a variety of shop equipment, including lathes, milling machines, drill presses, saws, grinders, a metal sheer and breaker, hand tools, and various power tools (Fig. 6.2-3). Additional instruction is given for the use computer-aided fabrication machines for more complicated parts: namely our Southwestern Industries CNC milling machines (a TRAK KE 2-Axis and a TRAK DMPSX2P 3-Axis), along with our KERN HSE large format laser cutter system (also purchased with UW Student Technology Fee funds). In total, five students and eight faculty/staff members were trained in the student shop this annual reporting period.

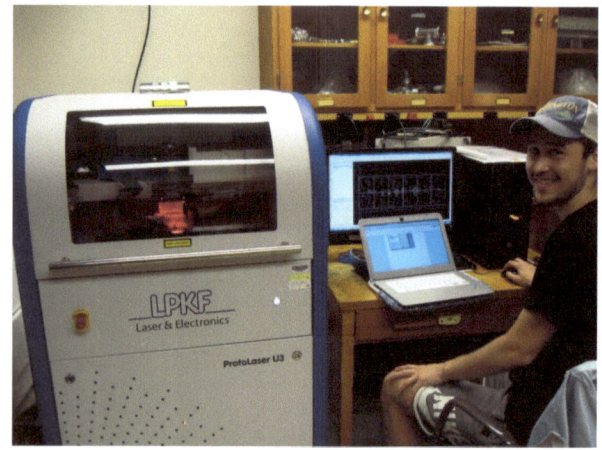

Figure 6.2-2. Undergraduate student Cedric Kong using CENPA's UV ProtoLaser for prototype PCB fabrication.

Figure 6.2-3. Undergraduate student Samuel Sexton using CENPA's TRAK KE 2-Axis CNC milling machine.

The long-standing unique history of teaching students how to operate our accelerator and ion sources remains a vital part of CENPA today. Graduate and undergraduate students receive instruction in the theory and practicalities of running our duoplasmatron ion source and 9-MegaVolt FN Tandem Van de Graaff accelerator (Fig. 6.2-4). This training incorporates both laboratory technical skills and safety through direct hands-on operation of the accelerator's equipment. Students practice generating an ion beam, charging the Van de Graaff terminal to multi-MegaVolt potentials, tuning the ion beam through the accelerator, and transporting the beam to an experimental target. After receiving this accelerator operations training (a.k.a. "crew training"), these students are qualified to operate the ion source and accelerator for their research. They may also be called upon to serve as crew operators during 24-hour accelerator operations for other research experiments.

Figure 6.2-4. Michael Huehn operating the tandem accelerator from the CENPA control room.

6.3 Accelerator-based lab class in nuclear physics

A. García, J. R. Pedersen, E. B. Smith, and D. I. Will

To provide accessible, hands-on education for working professionals, we have developed an evening graduate-level lecture and laboratory class to bolster nuclear physics knowledge in our local workforce[1] (Fig. 6.3-1). Focusing on relevant nuclear theory and practical techniques in nuclear physics experimentation, this class supplies both professional development for the growing aerospace and tech sectors, as well as a unique, hands-on learning environment for working professionals wishing to advance their careers.

---

[1]Phys 575, Nuclear Physics: Sources, detectors, and Safety, http://faculty.washington.edu/agarcia3/phys575

Figure 6.3-1. Home page for CENPA's Phys 575 class.

Each student received hours of experience using real nuclear physics equipment, utilizing our tandem Van de Graaff accelerator, an array of particle detectors, nuclear instrumentation electronics, and data acquisition systems (Fig. 6.3-2).

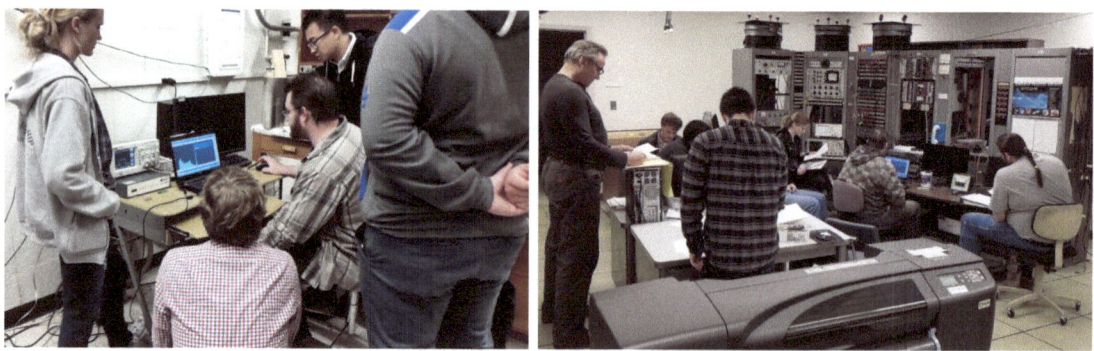

Figure 6.3-2. *Left*: Students viewing the gamma spectrum of a mystery radioactive source analyzed via a HPGe solid-state detector. *Right*: Students collecting RBS data from a 2 MeV proton beam scattering off of a mystery foil.

The class met twice a week during the Autumn 2017 quarter, once for a 1.5-hour lecture and again in groups for a 1.5-hour lab session. The subjects covered were:

1. Basics of nuclear physics, nuclear energy, orders of magnitude
2. Attenuation of photon radiation, solid-state detectors (Ge and Si)
3. Ranges of ions and electrons, the weak interaction, radioactivity, radiation damage, and health risks (α, β, γ, and neutron activity)
4. Deciphering a mystery γ spectrum measured using a Ge solid-state detector. Gauging the level of radioactivity and assessing health risks

5. 9 megavolt tandem accelerator function, duoplasmatron ion source function, and ion beam optics. Tuning beam through tandem accelerator
6. Measuring Rutherford back-scattering (RBS) spectra by detecting scattered protons from foil targets with Si solid-state detectors. Deciphering a mystery spectrum to determine the material contents of a mystery foil (Fig. 6.3-3 *Left*)
7. Fission and fusion. Basics of nuclear reactors
8. Nuclear astrophysics, nucleosynthesis in stars
9. Sources of positrons for positron emission tomography
10. Resonance energy of ^{27}Al(p,γ)^{28}Si nuclear reaction determined by irradiating an aluminum target with accelerated protons and detecting the resulting γ's in a high purity germanium (HPGe) solid-state detector (Fig. 6.3-3 *Right*)

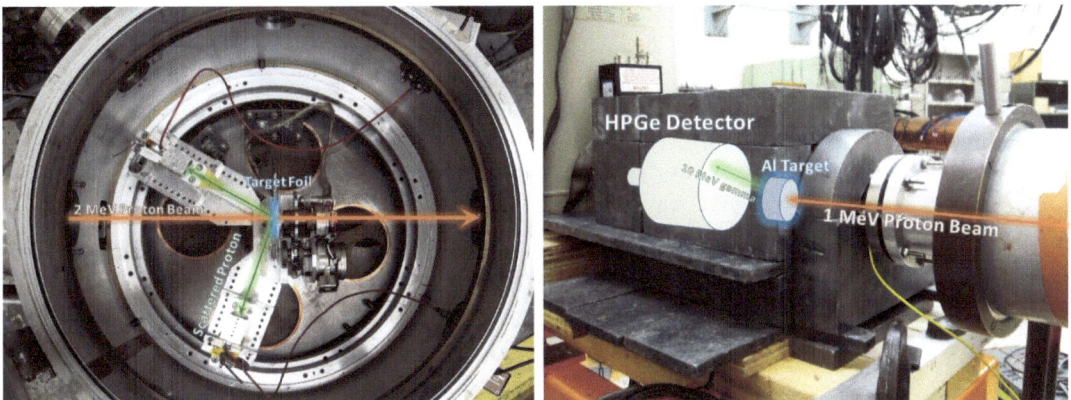

Figure 6.3-3. *Left*: RBS setup indicating 2 MeV protons (orange) incident on a thin foil target yielding back-scattered protons (green) detected in Si solid-state detectors. *Right*: ^{27}Al(p,γ)^{28}Si nuclear resonance setup indicating 1 MeV incident protons (orange) incident on Aluminum target yielding in resonance emission γ's (green) detected in a high purity germanium solid-state detector.

Thirteen students attended the class, where they engaged in enthusiastic discussions during lectures and in laboratory sessions. A class tour of the Boeing Radiation Effects Laboratory in Seattle was arranged by a former CENPA graduate student, who is now an employee of that facility.

7 Facilities

7.1 Laboratory safety

G. T. Holman, J. R. Pedersen, E. B. Smith, and D. I. Will

The culture and environment of safety at CENPA continues to improve. Achievements have been made in numerous areas, most notably in our UW Laboratory Safety Initiative safety survey score improvements and in the disposal of large quantities of chemical wastes and unused chemicals. Strides in safety training have been made in areas of providing, documenting, and compliance. As a combined effect of the attention and effort put toward laboratory safety, the safety survey score for the CENPA laboratories that are a part of the UW Laboratory Safety Initiative (LSI) has increased from 63 to 81 out of a possible 100 (81 is the current average for all UW Physics Department Laboratories).

UW laboratory safety initiative innovation event

CENPA was asked to present at the "Provost's Laboratory Safety Initiative Innovation Event" on December 11, 2017, to demonstrate our progress toward a culture of safety. Our presentation, "Center for Experimental Nuclear Physics and Astrophysics (CENPA) Safety Innovations", detailed the Safety Kiosk installations[1], associated Safety Kiosk webpage, and pseudocode development of a programmatical query of EH&S training records to facilitate automated, up-to-the-minute CENPA personnel training statuses.

CENPA safety orientation procedure

This year we also implemented a workplace safety orientation protocol. Now, all personnel at CENPA must complete the 'New Employee Safety Orientation' process. This process requires the supervisor to complete a tour of the facility with the new employee. The new employee is trained in emergency evacuation, reporting, and alarms; fire extinguisher, first aid, and exit locations; potential hazards; and personal protection equipment (PPE) requirements. In addition to the facility tour, the employee is required to complete UW EH&S safety courses. Once the supervisor and employee complete the orientation packet, each sign and date the packet and the completed form is saved in the employees personnel folder. To ensure safety training compliance, the employee is provided a door code for access to facility if and only if they have completed all necessary safety courses and they and their supervisor have completed the safety orientation packet.

Student shop safety

Analogous to the aforementioned CENPA safety orientation procedure, we also improved student shop safety awareness and training records. Prior to using the the student shop, employees are now required to meet with CENPA instrument maker David Hyde to review the employee training records and safety guidelines. Each employee now reviews and signs the student shop orientation and training record. This document covers personal protection

---
[1] CENPA Annual Report, University of Washington (2017) p. 145.

equipment (PPE) and safety guidelines. The document will also indicate which equipment employees are trained/permitted to use.

CENPA waste disposal initiative

This year we set the fairly aggressive goal of removing the majority of accumulated hazardous waste. The *CENPA Safety and Waste Disposal 2017 Initiative* outlined the following goals: assess the hazardous chemicals and waste in all labs/rooms; document the storage and containment status of identified hazardous materials; notify laboratory PIs of hazardous and unused materials planned for disposal; label containers of hazardous materials to be removed, and if necessary, transfer hazardous contents to suitable disposal containers; arrange for a disposal contractor to remove hazardous materials identified for disposal. On January 11th, 2018, with assistance from UW EH&S Matt Moeller and his team, 117 gallons of liquid hazardous waste and 127 pounds of solid hazardous waste were removed from CENPA (Fig. 7.1-3). The rooms we addressed: 123A (darkroom), 162 (hot lab), 163 (high-bay), 164 (courtyard), 165 (lead lab), 166 (cold lab) (see Fig. 7.1-1 and Fig. 7.1-2).

Figure 7.1-1. *Left images*: Before cleanup, *Right images*: After cleanup.

Figure 7.1-2. *Left images*: Before cleanup, *Right images*: After cleanup.

Figure 7.1-3. Hazardous waste in staged in barrels and containers awaiting pickup.

CENPA safety innovations

Progress was made on establishing and improving protocols for chemicals remaining at CENPA after the disposal initiative. A re-inventory of remaining chemicals and updating of MyChem (UW's online chemical inventory application), including assignment of room/lab responsible persons, was performed. Creating and ensuring the existence of accurate and updated standard operating procedures (SOPs) for all chemicals was aided by the reduced number of stored chemicals. A new protocol of assigning a unique serial number to each container of chemical in the laboratories (old and new) has been established. A high quality label printer has been linked to an online application that prints the serial number (numeric & QR coded), responsible person, and date acquired for each chemical container. This CENPA chemical serial number will be used in an in-house inventory application that will also link with MyChem. As part of the chemical acquisition protocol, an SOP is required before chemical purchases are made.

High-purity germanium (HPGe) gamma detector system

CENPA has worked to re-assemble and upgrade a high-purity germanium (HPGe) gamma detector system for the spectroscopic identification and activity survey of various known and unknown radioactive material (RAM) for inventory and disposal. Room 123A (previously a film dark room) has been cleaned and painted. It now contains the HPGe detector, associated nuclear instrumentation modules (NIM) & racks, and data acquisition (DAQ) equipment. CENPA has been awarded a Student Technology Fund (STF) grant to upgrade the NIM and DAQ for improved employee training in the use of the equipment to aid our efforts in RAM inventory & disposal, as well as for use in their own research initiatives. Additionally, a separate dedicated portable gamma spectroscopy system is being completed to facilitate the measurement of RAM that cannot be brought to RM 123A. This portable system will allow CENPA to make strides toward removing decades-old RAM from our site.

Improved compressed gas cylinder storage

The removal of waste from room 164 and subsequent pressure washing facilitates the use of the room as a new storage location for compressed gas cylinders. The room is a courtyard with roof eaves on all four walls, and by placing two 8-cylinder gas racks along one wall and a few wall mounted cylinder holders on an opposite wall, we will be able to move all full and empty compressed gas cylinders to this room. The two cylinder racks have been acquired and will be installed soon, eliminating the current and less desirable location along a steel pipe railing in the student shop.

7.2 Van de Graaff accelerator and ion-source operations and development

M. J. Borusinski[*], C. Cosby[†], H. T. Gorell, B. E. Hamre[‡], G. H. Leum, J. E. Oppor[§], J. R. Pedersen, D. A. Peterson, S. S. Sexton, E. B. Smith, S. B. Troy, T. D. Van Wechel, and D. I. Will

The only tandem entry occurred from Aug. 20 - Oct. 20, 2017 in order to correct a failure to control the vertical (VERT) high-voltage (HV) Terminal Steerer. The subsequent troubleshooting, diagnosis, and temporary solution was very complex, with many inter-related issues. First, what was most directly determined to be the cause of the VERT Steerer control failure was a blown fuse (Littlefuse micro 273 125V 0.5A) on the TERMINAL COMPUTER Motor Control PCB3 board (+) motor drive signal (P03 PIN 2), which was replaced with a new fuse. Further investigation uncovered an additional failure to readback the value of the VERT Steerer, which was determined to be a failed channel of an ADC card.

Detailed inspection of all transient voltage suppression components of the TERMINAL COMPUTER signal interface boards revealed tranzorbs (1.5KE 33C) on the Motor Control PCB2 board measuring less than acceptable isolation values on many signals (PINS 1-8, 13, 16-18, 20), so they were replaced with new tranzorbs (1.5KE 33CA).

Next, an investigation into the motor drive of the VERT and HORIZ Steerer variacs was pursued based on observing slow and erratic adjustment speeds. Planetary gear reducers for VERT and HORIZ variac motor drives were disassembled, degreased of old tacky grease, and regreased with new White Lithium Grease (multi-purpose NLGI #2). The HORIZ gear reducer was mounted with new spring-tightened mounts resulting in less torque drag from reducer-to-variac shaft coupling mis-alignment. One of the HORIZ Steerer variac limit microswitches had failed (a modified V3L-139-D8), and was replaced with a new microswitch that matches the other limit microswitches (V3L-1101-D8). Further investigation revealed that the middle pin on the VERT motor control assembly board "PS" (AC output voltage to DC power supplies) connector was pulled out, and required re-insertion and soldering to

---

[*]Departed June, 2017.
[†]Arrived July, Departed August, 2017.
[‡]Departed July, 2017.
[§]Departed June, 2017.

make connection. This loose pin could also have played a roll in the inability to adjust the VERT Steerer.

A continued failure to readback the position of the VERT Steerer variac position uncovered a failed channel CH01 on the GROUP 3 ADC "C" board (S/N C2004) in DI0 box of the TERMINAL COMPUTER. A replacement GROUP 3 "C" board was optained (S/N C21029), but it was not able to be communicated with by the DIO fiber optic comm board (S/N PR030). So, the original "C" board (S/N C2004) was re-installed, and the VERT variac position signal was moved to TERMINAL COMPUTER signal interface board AI4 connector P3. This connector P3 is normally reserved for a readback of the stripper gas leak valve position, so as a result, the VERT Steerer variac position readback now appears in the stripper gas leak valve position field of the CSX program. This "work around" also requires setting of the "Y ADC 1" LOWER ALARM on the LE GORDO (GORDO 1) program to 0 volts in order to allow the VERT motor control signals to be active.

Lastly, intermittent fiber optic communications to and from the TERMINAL COMPUTER were solved by replacing the fiber optic connectors at the TERMINAL COMPUTER "fiber optic junction box" on the SEND (to TERMINAL COMPUTER) and RECEIVE fibers running the length of the Low Energy (LE) accelerator column (the SEND fiber was nearly cracked through completely). Additionally, the RECEIVE fiber from the LE bulkhead to the LE GORDO loop control card had to be replaced with a new, shorter (only 5.5 feet) fiber to achieve constant SEND and RECEIVE communtications. This indicates that all the fibers likely have marginal transmission due to radiation damage, especially those inside the TANDEM accelerator, and are in need of replacement.

The majority of effort on the Ion Sources DECK during this reporting period was focused on producing a suitable beam of ^{21}Ne$^+$ for implanting Ta targets for an upcoming ^{21}Ne(p,g) experiment. However, important achievements of a new filament holder for the duoplasmatron Direct Extraction Ion Source (DEIS), new DEIS filament curing chamber, a hall probe zeroing chamber, and installation of water cooling flow meters also occured.

The work toward increased ^{21}Ne$^+$ implant current (i.e. current on the implant target) continues with the goal of achieving the 100 nA at 30 keV currents of September of 2009. The highest ^{21}Ne$^+$ current reached thus far in 2018 is 22 nA at 30 keV, with 47 nA of ^{21}Ne$^+$ at 30 keV in the DECK IMAGE cup and 53 uA of non-mass-analyzed Ne$^+$ beam in the DECK OBJECT cup. The first attempt, on January 2017, to produce Ne$^+$ yielded 11 uA in the OBJECT cup. A disassembly and cleaning of the DEIS bottle, extraction plate, extraction aperture retaining ring, and FOCUS electrode was preformed. We also installed a new 35 mil diameter hole extraction aperture; a re-polished, previously used, all steel plasma bottle nose, newly re-bored to 126 mil diameter canal. Finally, we surveryed the alignment of the bottle nose and extraction aperture.

The next use, May 2017, of the DEIS to extract Ar$^+$, as a learning platform instead of the more expensive Ne, resulted in low OBJECT cup beam currents, BIAS voltage (i.e. extraction voltage) power supply (PS) currents that were too high, and sparking issues that culminated in the FOCUS electrode behaving as a short to the extraction plate (i.e. COMMON or

GND of the Ion Source DECK). Disassembly of the DEIS revealed flaking metal deposits on the FOCUS electrode side of the extraction aperture insert retaining ring. The flaking aligned with arcing "scortch" marks on the FOCUS electrode, indicating that the flakes from the metal deposits were shorting the extraction electrode to the FOCUS electrode. The extraction insert, with 35 mil diameter hole, also had much buildup of deposited metal on its beam entrance side. Microscopic inspection of the all-steel plasma bottle nose showed distinct "barreling" of the exit canal due to sputtering. This sputtering of the steel is very likely the source of the deposited metals discribed above. The DEIS, extraction plate, insert retaining ring, and FOCUS electrode were all cleaned to remove metal deposits. The extraction insert was replaced with new 35 mil diameter hole insert, and the bottle nose by a newly machined all-steel bottle nose (#5) with 120 mil diameter canal.

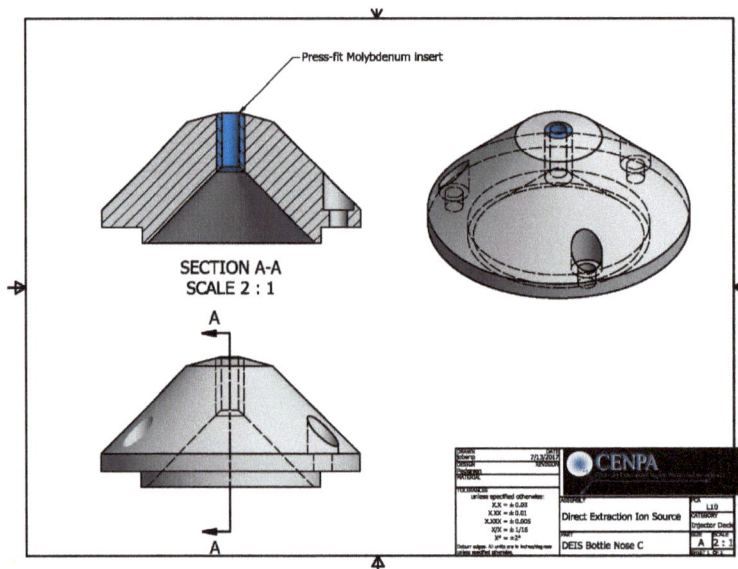

Figure 7.2-1. New DEIS plasma bottle nose design with 40 mil thick wall of Molybdenum lining canal to reduce sputtered removal of steel by extracted ions.

In July 2017, a new DEIS plasma bottle nose (#6) was designed and fabricated using Molybdenum (Mo) metal in the canal to reduce sputtering (Fig. 7.2-1). Bottle nose #6 had exactly the same cross-section as the previous versions, except for a 40 mil-thick wall of Mo lining the 120 mil-diameter exit canal. The electric field at the canal surface and in the gap between the bottle nose and extraction plate should be the same, since Mo is conductive like the steel, but the magnetic field in the canal and in the gap will be slightly less confining, since the Mo is non-magnetic. This Mo-walled canal bottle nose was installed, allowing for production of Ne^+ without any sparking or shorting of the FOCUS electrode. However, the Ne^+ current extracted was still very low, and the BIAS voltage PS current too high.

In January 2018, an all-steel DEIS plasma bottle nose (#3), which had been used in 2013 to produce a $^{36}Ar^+$ beam for implanting and used for standard negative ion extractions for several years, was polished and installed. Its canal was the standard 120 mil diameter, but it too showed signs of barreling (i.e. sputtering). Progressive tuning of Ne^+ and Ar^+

beams over the course of weeks lead to the 22 nA ^{21}Ne$^+$ beam at the implant target in April. This required precise x-y positioning of the DEIS bottle nose canal with respect to the 35 mil-diameter extraction aperture in order to maximize the ratio of extracted OBJECT cup current verses BIAS voltage PS current, and required development of a tool to allow for vertical adjustment of the DIES while at BIAS voltage. Additionally, BECKMAN model HV-211-22 high voltage (HV) probes were connected to the extraction plate and FOCUS electrode for the direct measure of these voltages without relying on the power supply panel meters, which were found to give inaccurate measurements at the given node. Confirming that the FOCUS electrode was at a different voltage than the extract voltage, thus producing a focused beam, was very important to the beam current optimization.

During this annual reporting period from April 1, 2017, to March 31, 2018, the tandem pellet chains operated 247 hours, the sputter-ion-source (SpIS) operated 0 hours, and the duoplasmatron direct-extraction-ion source (DEIS) operated 513 hours. Additional statistics for accelerator operations are given in Table 7.2-1. The tandem accelerator-produced beams of ions originating from the DEIS included 17.8 MeV ^{2}H$^+$ for the ^{6}He experiment, 0.99 to 2.0 MeV ^{1}H$^+$ for the Physics 575 class, and 15.2 MeV ^{2}H$^+$ for visiting experimenters.

ACTIVITY SCHEDULED	DAYS SCHEDULED	PERCENT of TIME
Nuclear physics research, accelerator	19	5.2
Development, maintenance, or crew training	103	28.2
Grand total	122	33.4

Table 7.2-1. Tandem Accelerator Operations from April 1, 2017 to March 31, 2018.

7.3 Laboratory computer systems

G. T. Holman

CENPA is a mixed shop of Windows 7, 10, Mac OS X, and various Linux distributions. Windows 10 is installed on new workstations, but we are still running some Windows XP systems for data acquisition, DOS 6 on accelerator controllers, and an embedded Win 98 for a mechanical shop mill. As with every year, the IT focus was directed toward server consolidation, network security, process documentation and removal of redundant processes. We continue to utilize Xen virtualization for Autodesk Vault versions, ELOGs, wikis, collaboration calendars, and document servers. The CENPA website and research group web pages run on an upgraded server and have been migrated to Drupal 7 web framework. The NPL mail server still provides NPL presence but all email is relayed to UW e-mail hardware. Workstations connect to the UW delegated organizational unit (OU), which mostly removes the need to run a dedicated domain or LDAP server.

Two Dell 510 20-TB servers (Lisa and Marie) continue to offer user storage, print server capability, and improved backup policy. Linux, Windows, and Mac workstations are backed up to the 20-TB Marie raid farm, which is backed up offsite using UW Trivoli backup service. Lisa runs the Crash Plan Pro backup application which supports all operating systems and provides differential and encrypted backups. Whereas workstations rely on Crash Plan Pro for backups, all servers utilize rsync. Marie provides 20 TB for research, user, and shared group data.

The NPL Data Center (NPLDC) provides legacy infrastructure supporting high-performance scientific-computing applications. The cluster comprises two specific cluster instances. The first cluster instance 'cenpa-mamba' runs the latest open-source Rocks software[1], runs the cvmfs (Cern-VM file system) client, and Frontier local squid cache server. The second cluster uses Rocks version 5.4, and most notably runs COMSOL, Cern ROOT, and Geant. Both cluster instances use Torque/Maui or Sun Grid Engine (SGE) via dedicated front ends. Approximately one third of the rack space is dedicated to non-cluster hardware: scratch storage, SQL, ELOGs, web applications, CAD workstations, and backup storage. These servers constitute over 200 TB of raw disk space.

To upgrade our existing cluster hardware, we acquired new servers from the National Energy Research Scientific Computing Center (NERSC). The donation included 105 Dell R410 (compute nodes), 5 R710 (raid controllers), 12 MD1200 (storage arrays). Storage arrays allow up to 12-4TB drives, total 48TB/each, total 768TB all filled. Projected cost to fill all drive bays is projected to $19.2k for the raid arrays and approximately $9k to upgrade gigabit to 10Gbe network backplane.

7.4 Electronic shop

D. A. Peterson and T. D. Van Wechel

The electronics shop is responsible for the design and construction of new laboratory electronic equipment as well as the maintenance and repair of existing CENPA electronics. Projects undertaken by the electronics shop in the past year include the following:

1. The design and construction of the $g-2$ NMR timing interface which provides a timing signal to the NMR digitizers and a fire pulse to the NMR NIM modules which is then passed to the multiplexers (Fig. 7.4-1).

---

[1] http://www.rocksclusters.org/

Figure 7.4-1. *Left*: $g-2$ NMR timing interface. *Right*: the DIN rail interface to the Acromag FPGA digital I/O board.

2. The design and construction of the $g-2$ Inflector Beam Montoring System (IBMS) for beamline positions one and two. IBMS position #1 was prototyped in 2016 and produced in 2017. IBMS position #2 was prototyped and produced in 2017; its main feature was to utilize an edge connector to interface to boards with varying SiPM pitches and to simplify electronics replacement (Fig. 7.4-2). The PCB for IBMS #3 will use the same printed circuit boards as IBMS #2.

Figure 7.4-2. *Top*: IBMS #1, *Bottom*: IBMS #2 with plug-in boards for both axes.

3. The construction of four IBMS breakout boards. The boards interface digital control of the IBMS perhiperals, low voltage, bias voltage, and distribute input test pulses. It is also a pass-through for the output signals from the IBMS signal cable bundle to

the digitizers through micro coaxial connectors (Fig. 7.4-3).

Figure 7.4-3. *Left*: IBMS breakout front panel. *Right*: IBMS breakout back panel.

4. The design and construction of a LED pulser for the $g-2$ IBMS calibration (Fig. 7.4-4).

Figure 7.4-4. IBMS four-channel LED pulser.

5. The design and construction of new photomultiplier tube (PMT) bases for the COHERENT experiment. The electronics shop will assemble approximately 110 of these boards this summer (Fig. 7.4-5).

Figure 7.4-5. COHERENT PMT base.

6. The construction of the $g-2$ high-voltage (HV) probe to Cornell digitizer adapter (Fig. 7.4-6).

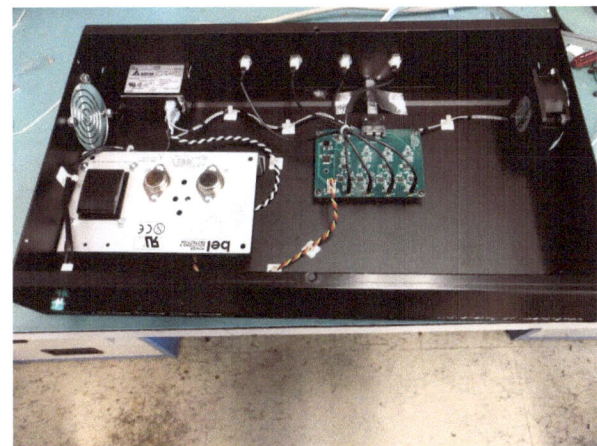

Figure 7.4-6. $g-2$ high-voltage (HV) probe to Cornell digitizer interface.

7.5 Instrument Shop

T. H. Burritt, J. H. Elms, D. R. Hyde, S. Kimes, and H. Simons

The CENPA instrument shop has provided design and fabrication support to CENPA and several UW research groups. Projects completed this year include:

1. TRIMS bellows adjustment with micrometer head.

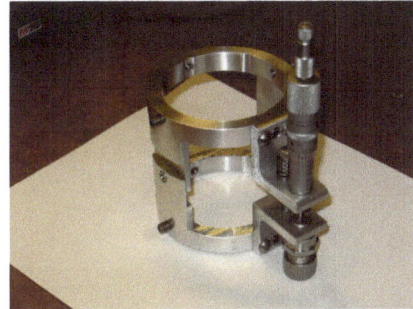

Figure 7.5-1. Device is used to sweep and assist with alignment.

2. High 80-amp feed-through to fix problem with magnets over heating.

Figure 7.5-3. *Top left:* KATRIN IRON BIRD vacuum and cold-head feed. *Top right:* Thin heat shield inside vacuum. *Bottom:* High amperage feed-through.

3. MAJORANA infrared copper shield.

Figure 7.5-4. Thin wall welded copper.

4. KATRIN split bellows support with c-ring and vacuum adapter.

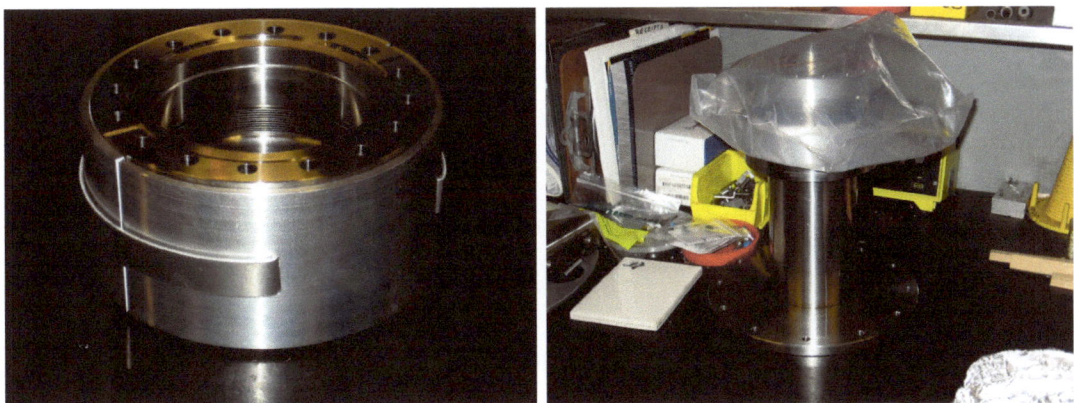

Figure 7.5-6. *Left:* Support for vacuum and easy axis to bolts. *Right:* Bellows vacuum adapter flange to detector.

5. $g-2$ Double photo tube, SiPM and filter scintillator.

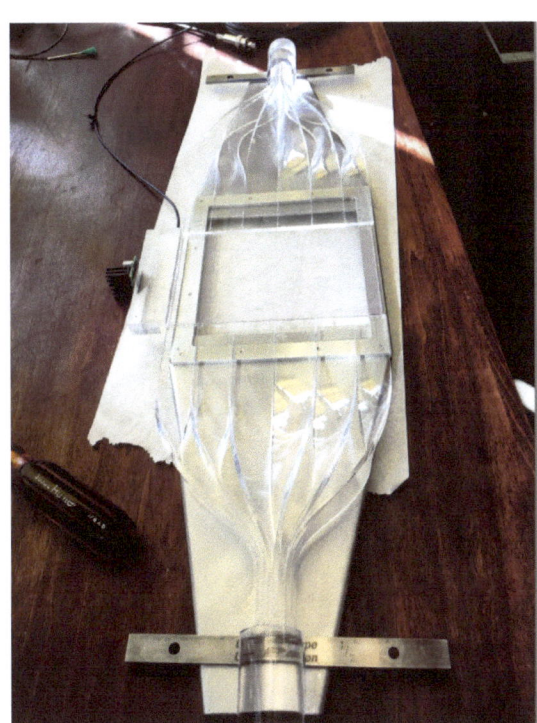

Figure 7.5-7. $g-2$ completed scintillator assembly.

6. Gravity

Figure 7.5-8. *Left:* GRAVITY 10" diameter assembly. *Right:*: GRAVITY balance for assemblies.

7. DAMIC removal and rebuild

Figure 7.5-10. DAMIC Clean room removal from B037 and rebuilt in B059 with new floor plan.

8. Project 8 winding coil and wave guide parts.

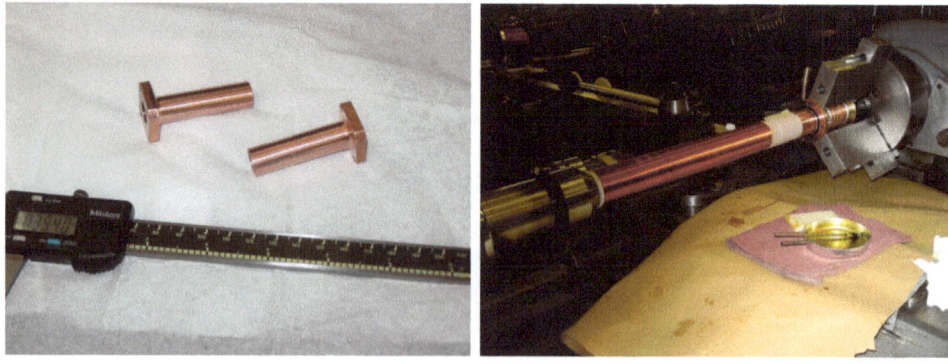

Figure 7.5-12. Project 8 Wave guide copper parts and winding copper coil set up in shop.

9. Project 8 electrical box

Figure 7.5-13. Project 8 Machine electrical box for components.

10. PCA junction boxes

Figure 7.5-14. PCA Junction boxes fabrication one of many all shapes and sizes.

11. Fabricated new and repaired components for Van De Graaff accelerator materials using multiple materials (e.g. stainless, titanium, copper, mumetal, tantalum, nickel, maycore, tungsten vespel, and peek) (Sec. 7.2).

7.6 Building maintenance, repairs, and upgrades

J. Pedersen, E. B. Smith, and D. I. Will

During the 2017 calendar year 173 Work Orders (WOs) were placed for the North Physics Laboratory buildings which house the CENPA research facility. For the Van de Graaff Accelerator Building 75 total WOs included the following: 42 Preventive Maintenance/Repair/Alterations WOs (PM/R/Alts) by Facilities Services; 25 Service/Repair WOs (S/R) by CENPA staff; and 8 Alterations WOs (Alts) by CENPA staff. For the Cyclotron Building 59 total WOs included the following: 40 PM/R/Alts WOs by Facilities Services, 16 S/R WOs by CENPA staff, and 3 Alts WOs requested by CENPA staff. For the CENPA Instrument Shop 39 total WOs included the following: 28 PM/R/Alts WOs by Facilities Services, 10 S/R WOs by CENPA staff, and 1 Alts WOs by CENPA staff.

In particular, one direct steam heat air handler in the Cyclotron Building began leaking hot water from a site near the condensate end of its heat exchanger. A new heat exchanger was purchased and installed by UW Facilities Services personnel.

This past winter the set-up area high bay (room 163) was particularly cold. A vent fan in this area was found to have stuck louvers which failed to stop inflow of cold air. Those louvers are being replaced. Two direct steam heat units in this same area were found to have failed, partly because the steam had been turned off for leak repairs, and also due to an electrical motor failure. These have been repaired. We still await replacement of a failed condensate trap, re-lagging of the steam and condensate lines, and vent louver replacement.

For some years now, we have struggled with water hammer in the hydronic heating system for the Van de Graaff Building. Lab personnel working with the UW controls shop determined the cause was persistent over-pressurization of this hot water system. A fix has been designed and approved. A debris strainer will be added and the leaking make-up regulator and a failed pressure gauge will be replaced. To prevent over-pressurizations in the future, a parallel over-pressure relief (with lower setting capability than the required safety) will be added.

During fall 2017, with the help of UW Radiation Safety and collaborators at Pacific Northwest National Laboratory (PNNL), 3 tons of depleted uranium gravity masses were disposed of properly via transfer to the PNNL collaborators.

The old darkroom and circuit etching lab, room 123, has been cleared for reuse. It is a potential location for our general-lab-use High Purity Germanium (HPGe) detector system (see (Sec. 7.1)). Evaluation of temperature stability in this room is ongoing.

Finally, staff and students continue removing and disposing of remaining portions of the decommissioned CENPA superconducting LINAC booster and its infrastructure to create space for the Orpheus Project (Sec. 5.2) at the west end of our accelerator tunnel. During this reporting period, 1.5 tons of steel I-beams, 10.5 tons of steel shielding, 9 tons of concrete pillars, and several hundred feet of wires and cables were removed and recycled.

Two worn out Koch RS Rotary Screw Helium Compressors and all eight Beech Aircraft, Boulder Division, cryogenic valve boxes with lines have been disposed of via UW Surplus.

8 CENPA Personnel

8.1 Faculty

Eric G. Adelberger[1]	Professor Emeritus
Hans Bichsel[1]	Affiliate Professor
Alvaro Chavarria[1,2]	Assistant Professor
John G. Cramer[1]	Professor Emeritus
Jason Detwiler	Assistant Professor
Peter J. Doe	Research Professor
Sanshiro Enomoto	Research Assistant Professor
Martin Fertl	Research Assistant Professor
Alejandro García	Professor
Gerald Garvey[1]	Affiliate Professor
Jens H. Gundlach[1]	Professor
Blayne R. Heckel[1]	Professor; Chair
David W. Hertzog	Professor; Director
C. D. Hoyle[1,3]	Affiliate Assistant Professor
Peter Kammel	Research Professor
Jarek Kaspar	Research Assistant Professor
Michael L. Miller[1]	Affiliate Research Assistant Professor
Peter Mueller[1]	Affiliate Professor
Diana Parno[1,4]	Affiliate Assistant Professor
R. G. Hamish Robertson	Emeritus Professor
Leslie J Rosenberg[1]	Professor
Gray Rybka[1]	Research Assistant Professor
Kurt A. Snover[1]	Research Professor Emeritus
Derek W. Storm[1]	Research Professor Emeritus
Thomas A. Trainor[1]	Research Professor Emeritus
Robert Vandenbosch[1]	Professor Emeritus
Krishna Venkateswara[1]	Acting Assistant Professor
William G. Weitkamp[1]	Research Professor Emeritus
John F. Wilkerson[1,5]	Affiliate Professor

8.2 CENPA external advisory committee

Robert McKeown[6]	Jefferson Laboratory
Daniel McKinsey[6]	UC Berkeley
Michael Ramsey-Musolf[6]	University of Massachusetts, Amherst

---

[1]Not supported by DOE CENPA grant.
[2]Arrived July, 2017.
[3]Affiliated faculty, Humboldt State University, Arcata, CA.
[4]Affiliated faculty, Carnegie Mellon University, Pittsburgh, PA.
[5]Affiliated faculty, University of North Carolina, Chapel Hill, NC.
[6]CENPA External Advisory Committee formed January 2014.

8.3 Postdoctoral research associates

Brent Graner[1]
Mathieu Guigue[0,2]
Charlie Hagedorn[0]
Megan Ivory[5]
Rakshya Khatiwada[0]
Kim Siang Khaw

Benjamin LaRoque[0,2]
Elise Novitski[3]
Richard Ottens[0,4]
Walter Pettus
Daniel Salvat

8.4 Predoctoral research associates

Sebastian Alvis[6]
Yelena Bagdasarova
Hannah Binney[6]
Micah Buuck
William Byron[0,8]
Raphael Cervantes[0]
Nick Du[0]
Ali Ashtari Esfahani[0]
Aaron Fienberg
Nathan Froemming
Julieta Gruszko[10]
Ian Guinn
Jason Hempstead
Luke Kippenbrock

John Lee[0]
Erik Lentz[0,7]
Ying-Ting Lin
Brynn MacCoy
Eric Machado
Eric Martin[9]
Ethan Muldoon
Rachel Osofsky
Michael Ross[0,6]
Nicholas Ruof[6]
Rachel Ryan
Erik Shaw[0]
Matthias Smith[10]
Matthew Turner[11]

---

[1] Graduated August, 2017. Accepted postdoc position.
[2] Pacific Northwest National Laboratory, Richland, WA.
[3] Arrived September, 2017.
[4] Departed August, 2017.
[5] Arrived January, 2018.
[6] Arrived June, 2017.
[7] Graduated October, 2017.
[8] Arrived September, 2017.
[9] Graduated May, 2017.
[10] Graduated August, 2017.
[11] Graduated November, 2017.

8.5 Undergraduates

Robert Adams	Rybka, Advisor
Yifei Bai	Hagedorn, Advisor
Tanner Bambrick	Rybka, Advisor
Elizabeth Bernbaum	Rybka, Advisor
Carson Blinn	Hagedorn, Advisor
Jeff Capoeman, Jr	Hagedorn, Advisor
Jameson Doane	Rybka, Advisor
Ana Duarte	Detwiler, Advisor
Katherine Evans	Smith, Advisor
Roland Farrell	García, Advisor
Sean Fell	Rybka, Advisor
Sandor Fogassy	Detwiler, Advisor
Kezhu Guo	García, Advisor
Joshua Handjojo	Fertl, Advisor
Patrick Harrington	Rybka, Advisor
Nils Hostage	García, Advisor
Brandon Iritani	Hagedorn, Advisor
Jacob Johnson	Rybka, Advisor
Jared Johnson	Hagedorn, Advisor
Tyler La Rochelle	García, Advisor
Alyssa Lee	Rybka, Advisor
Connor Leupold	Venkateswara, Advisor
Jocelin Liteanu	Detwiler, Advisor
Jeremy Lu	García, Advisor
Tia Martineau[1]	Rybka, Advisor
Cayenne Matt	Hagedorn, Advisor
Parashar Mohapatra	Rybka, Advisor
Axl O'Neal	Smith, Advisor
Nicholas Orndorff	Hagedorn, Advisor
Maurice Ottiger	Fertl, Advisor
Yujin Park	Rybka, Advisor
Benjamin Phillips	Rybka, Advisor
Elliott Phillips	Fertl, Advisor
Daniel Primosch	Rybka, Advisor
Spencer Pruitt	Detwiler, Advisor
Hava Schwartz[2]	Detwiler, Advisor
Jennifer Smith[3]	Rosenberg/Rybka, Advisor
Matthew Stortini	García, Advisor
MinJung Sung	García, Advisor

---

[1] 2017 REU student, UMass Dartmouth.
[2] 2017 REU student, Harvey Mudd College.
[3] 2017 REU student, Stanford University.

Sadhana Suresh[1]   García, Advisor
Xinmiao Tang   Chavarria, Advisor
Andrew Thornberry   Rybka, Advisor
Colin Watson   Detwiler, Advisor
Kassandra Weber   Rybka, Advisor
Zachary Wuthrich   Detwiler, Advisor
Anni Xiong   García, Advisor

8.6 Visitors and volunteers

Alessandro Bandocci   Visiting Scientist, Project 8
Christian Boutan   Visitor, ADMX
Daniel Bowring   Visitor, ADMX
Thomas Braine   Volunteer RA, ADMX
Raahul Buch   Volunteer RA, He6
Nicholas Buzinsky   Visiting RA, Project 8
Alexander Cable   Visitor, LPKF laser system user
Jared Canright   Volunteer RA, KATRIN
Aaron Chou   Visitor, ADMX
Christine Claessens   Visiting RA, Project 8
Akash Dixit   Visiting RA, ADMX
Laura Gladstone   Visiting Postdoc
Yiran Fu   Visitor, ADMX
William Holden   Visitor, LPKF laser system user
Alexandru Hostiuc   Volunteer RA, LEGEND/Majorana
Aamar Ieso   Visiting Research Scientist, ADMX
Evan Jahrman   Visitor, LPKF laser system user
Scott Kihara   Visitor, LPKF laser system user
Cedric Kong   Visitor, LPKF laser system user
Pitam Mitra   Visiting RA, DAMIC
Bernadette Maria Rebeiro   Visiting RA, Ne21pg
Manuja Sharma   Visitor, LPKF laser system user
Daniel Simons   Volunteer, ADMX
Yu-Hao Sun   Visiting RA, Project 8
David Tanner   Visitor, ADMX
Smarajit Triambak   Visitor, Beamline Project
Francis Walsh   Volunteer RA, He6
Nathan Woollett   Visitor, ADMX
Jihee Yang   Visiting Research Associate, ADMX
David Young   Visiting RA, Kern laser system user

---

[1] 2017 REU student, U of Connecticut.

8.7 Professional staff

John F. Amsbaugh	Research Engineer	Engineering, vacuum, cryogenics design
Tom H. Burritt	Shop Supervisor	Precision design, machining
Nick Force[1]	Research Engineer	ADMX
Gary T. Holman	Associate Director	Computer systems
Seth Kimes[2]	Research Engineer	ADMX Dilfridge
Joben Pedersen	Research Engineer	Accelerator, ion sources
Duncan J. Prindle, Ph.D.	Research Scientist	Heavy ion, muon research
Eric B. Smith	Research Engineer	Accelerator, ion sources
H. Erik Swanson, Ph.D.	Research Physicist	Precision experimental equipment
Timothy D. Van Wechel	Research Engineer	Analog and digital electronics design
Douglas I. Will	Senior Engineer	Cryogenics, ion sources, buildings

8.8 Technical staff

James H. Elms	Instrument Maker
David R. Hyde[3]	Instrument Maker
Matthew Kallander[4]	Laboratory Technician
David A. Peterson	Electronics Technician

8.9 Administrative staff

Ida Boeckstiegel	Office Administrator
Victoria A. Clarkson[5]	Administrator
Wing Lam (Celia) Chor[6]	Fiscal Specialist
Nerissa Pineda[7]	Fiscal Specialist
Brice Ritter[8]	Fiscal Specialist

---

[1] Promoted to Research Engineer 2 in August, 2017.
[2] Job reclassification from classified to professional staff on December, 2017.
[3] Retired July, 2017. Accepted student-shop manager position September, 2017.
[4] Working on TRIMS starting August, 2017.
[5] Passed away after a brief, courageous battle with cancer on 4/27/17.
[6] Arrived April, 2017. Departed April, 2018.
[7] Arrived March, 2018. Departed June, 2018.
[8] Arrived August, 2017. Departed November, 2017.

8.10 Part-time staff and student helpers

Michael Borusinski[1]	Michael Huehn[2]
Christopher Cosby[3]	Tyler Larochelle[4]
Eric Erkela[5]	Grant H. Leum
Zhenghao (Mo) Fu[6]	Joshua Oppor[1]
Daniel Garratt[2]	Frederik Otto[7]
Henry Gorrell	Samuel Sexton
Brett Hamre[8]	Hendrik Simons
Greg Harper[9]	Santos Zaid[5]

---

[1] Departed June, 2017.
[2] Arrived November, 2017.
[3] Arrived July, 2017. Departed August, 2017.
[4] Arrived October, 2017. Departed March, 2018.
[5] Arrived June, 2017.
[6] Arrived January, 2017. Departed May, 2017.
[7] Arrived June, 2017. Departed July, 2017.
[8] Departed July, 2017.
[9] Arrived December, 2017.

9 Publications

Publications and presentations with a date of (2017) or (2018) are included below. Some entries from early (2018) may therefore also appear in the (2017) Annual Report.

9.1 Published papers

[1] N. Du, N. Force, R. Khatiwada, E. Lentz, R. Ottens, L. J Rosenberg, G. Rybka, G. Carosi, N. Woollett, D. Bowring, A. S. Chou, A. Sonnenschein, W. Wester, C. Boutan, N. S. Oblath, R. Bradley, E. J. Daw, A. V. Dixit, J. Clarke, S. R. O'Kelley, N. Crisosto, J. R. Gleason, S. Jois, P. Sikivie, I. Stern, N. S. Sullivan, D. B Tanner, and G. C. Hilton, "Search for invisible axion dark matter with the axion dark matter experiment", Phys. Rev. Lett. **120**, 151301 (2018).

[2] C. E. Aalseth et al., "Search for Neutrinoless Double-β Decay in ^{76}Ge with the Majorana Demonstrator", Physical Review Letters **120**, 132502, 132502 (2018), DOE Supported.

[3] N. Abgrall, I. J. Arnquist, F. T. Avignone III, A. S. Barabash, F. E. Bertrand, A. W. Bradley, V. Brudanin, M. Busch, M. Buuck, J. Caja, M. Caja, T. S. Caldwell, C. D. Christofferson, P.-H. Chu, C. Cuesta, J. A. Detwiler, C. Dunagan, D. T. Dunstan, Y. Efremenko, H. Ejiri, S. R. Elliott, T. Gilliss, G. K. Giovanetti, J. Goett, M. P. Green, J. Gruszko, I. S. Guinn, V. E. Guiseppe, C. R. S. Haufe, R. Henning, E. W. Hoppe, B. R. Jasinski, M. F. Kidd, S. I. Konovalov, R. T. Kouzes, A. M. Lopez, J. MacMullin, R. D. Martin, R. Massarczyk, S. J. Meijer, S. Mertens, J. H. Meyer, J. Myslik, C. O'Shaughnessy, A. W. P. Poon, D. C. Radford, J. Rager, A. L. Reine, J. A. Reising, K. Rielage, R. G. H. Robertson, B. Shanks, M. Shirchenko, A. M. Suriano, D. Tedeschi, L. M. Toth, J. E. Trimble, R. L. Varner, S. Vasilyev, K. Vetter, K. Vorren, B. R. White, J. F. Wilkerson, C. Wiseman, W. Xu, E. Yakushev, C.-H. Yu, V. Yumatov, I. Zhitnikov, and B. X. Zhu, "The processing of enriched germanium for the MAJORANA DEMONSTRATOR and R&D for a next generation double-beta decay experiment", Nuclear Instruments and Methods in Physics Research A **877**, 314–322 (2018), DOE Supported.

[4] B. Aharmim et al., "Search for neutron-antineutron oscillations at the Sudbury Neutrino Observatory", Phys. Rev. D **96**, 092005, 092005 (2017), DOE Supported.

[5] N. Abgrall, I. J. Arnquist, F. T. Avignone III, A. S. Barabash, F. E. Bertrand, M. Boswell, A. W. Bradley, V. Brudanin, M. Busch, M. Buuck, T. S. Caldwell, C. D. Christofferson, P.-H. Chu, C. Cuesta, J. A. Detwiler, C. Dunagan, Y. Efremenko, H. Ejiri, S. R. Elliott, Z. Fu, V. M. Gehman, T. Gilliss, G. K. Giovanetti, J. Goett, M. P. Green, J. Gruszko, I. S. Guinn, V. E. Guiseppe, C. R. Haufe, R. Henning, E. W. Hoppe, M. A. Howe, B. R. Jasinski, K. J. Keeter, M. F. Kidd, S. I. Konovalov, R. T. Kouzes, A. M. Lopez, J. MacMullin, R. D. Martin, R. Massarczyk, S. J. Meijer, S. Mertens, J. L. Orrell, C. O'Shaughnessy, A. W. P. Poon, D. C. Radford, J. Rager, A. L. Reine, K. Rielage, R. G. H. Robertson, B. Shanks, M. Shirchenko, A. M. Suriano, D. Tedeschi, J. E. Trimble, R. L. Varner, S. Vasilyev, K. Vetter, K. Vorren, B. R. White, J. F. Wilkerson, C. Wiseman, W. Xu, C.-H. Yu, V. Yumatov, I. Zhitnikov, and B. X. Zhu,

"The MAJORANA DEMONSTRATOR calibration system", Nuclear Instruments and Methods in Physics Research A **872**, 16–22 (2017), DOE Supported.

[6]COHERENT Collaboration, D. Akimov, J. B. Albert, P. An, C. Awe, P. S. Barbeau, B. Becker, V. Belov, A. Brown, A. Bolozdynya, B. Cabrera-Palmer, M. Cervantes, J. I. Collar, R. J. Cooper, R. L. Cooper, C. Cuesta, D. J. Dean, J. A. Detwiler, A. Eberhardt, Y. Efremenko, S. R. Elliott, E. M. Erkela, L. Fabris, M. Febbraro, N. E. Fields, W. Fox, Z. Fu, A. Galindo-Uribarri, M. P. Green, M. Hai, M. R. Heath, S. Hedges, D. Hornback, T. W. Hossbach, E. B. Iverson, L. J. Kaufman, S. Ki, S. R. Klein, A. Khromov, A. Konovalov, M. Kremer, A. Kumpan, C. Leadbetter, L. Li, W. Lu, K. Mann, D. M. Markoff, K. Miller, H. Moreno, P. E. Mueller, J. Newby, J. L. Orrell, C. T. Overman, D. S. Parno, S. Penttila, G. Perumpilly, H. Ray, J. Raybern, D. Reyna, G. C. Rich, D. Rimal, D. Rudik, K. Scholberg, B. J. Scholz, G. Sinev, W. M. Snow, V. Sosnovtsev, A. Shakirov, S. Suchyta, B. Suh, R. Tayloe, R. T. Thornton, I. Tolstukhin, J. Vanderwerp, R. L. Varner, C. J. Virtue, Z. Wan, J. Yoo, C.-H. Yu, A. Zawada, J. Zettlemoyer, A. M. Zderic, and aff13, "Observation of coherent elastic neutrino-nucleus scattering", Science **357**, 1123–1126 (2017), DOE Supported.

[7]M. Agostini, G. Benato, and J. A. Detwiler, "Discovery probability of next-generation neutrinoless double-β decay experiments", Phys. Rev. D **96**, 053001, 053001 (2017), DOE Supported.

[8]N. Abgrall, E. Aguayo, F. T. Avignone, A. S. Barabash, F. E. Bertrand, A. W. Bradley, V. Brudanin, M. Busch, M. Buuck, D. Byram, A. S. Caldwell, Y.-D. Chan, C. D. Christofferson, P.-H. Chu, C. Cuesta, J. A. Detwiler, C. Dunagan, Y. Efremenko, H. Ejiri, S. R. Elliott, A. Galindo-Uribarri, T. Gilliss, G. K. Giovanetti, J. Goett, M. P. Green, J. Gruszko, I. S. Guinn, V. E. Guiseppe, R. Henning, E. W. Hoppe, S. Howard, M. A. Howe, B. R. Jasinski, K. J. Keeter, M. F. Kidd, S. I. Konovalov, R. T. Kouzes, B. D. LaFerriere, J. Leon, A. M. Lopez, J. MacMullin, R. D. Martin, R. Massarczyk, S. J. Meijer, S. Mertens, J. L. Orrell, C. O'Shaughnessy, N. R. Overman, A. W. P. Poon, D. C. Radford, J. Rager, K. Rielage, R. G. H. Robertson, E. Romero-Romero, M. C. Ronquest, C. Schmitt, B. Shanks, M. Shirchenko, N. Snyder, A. M. Suriano, D. Tedeschi, J. E. Trimble, R. L. Varner, S. Vasilyev, K. Vetter, K. Vorren, B. R. White, J. F. Wilkerson, C. Wiseman, W. Xu, E. Yakushev, C.-H. Yu, V. Yumatov, and I. Zhitnikov, "Muon flux measurements at the davis campus of the sanford underground research facility with the MAJORANA DEMONSTRATOR veto system", Astroparticle Physics **93**, 70–75 (2017), DOE Supported.

[9]N. Abgrall, I. J. Arnquist, F. T. Avignone, A. S. Barabash, F. E. Bertrand, A. W. Bradley, V. Brudanin, M. Busch, M. Buuck, T. S. Caldwell, Y.-D. Chan, C. D. Christofferson, P.-H. Chu, C. Cuesta, J. A. Detwiler, C. Dunagan, Y. Efremenko, H. Ejiri, S. R. Elliott, T. Gilliss, G. K. Giovanetti, J. Goett, M. P. Green, J. Gruszko, I. S. Guinn, V. E. Guiseppe, C. R. S. Haufe, R. Henning, E. W. Hoppe, S. Howard, M. A. Howe, B. R. Jasinski, K. J. Keeter, M. F. Kidd, S. I. Konovalov, R. T. Kouzes, A. M. Lopez, J. MacMullin, R. D. Martin, R. Massarczyk, S. J. Meijer, S. Mertens, C. O'Shaughnessy, A. W. P. Poon, D. C. Radford, J. Rager, A. L. Reine, K. Rielage, R. G. H. Robertson, B. Shanks, M. Shirchenko, A. M. Suriano, D. Tedeschi, J. E. Trimble, R. L. Varner, S. Vasilyev, K. Vetter, K. Vorren, B. R. White,

J. F. Wilkerson, C. Wiseman, W. Xu, E. Yakushev, C.-H. Yu, V. Yumatov, I. Zhitnikov, B. X. Zhu, and MAJORANA Collaboration, "New Limits on Bosonic Dark Matter, Solar Axions, Pauli Exclusion Principle Violation, and Electron Decay from the Majorana Demonstrator", Physical Review Letters **118**, 161801, 161801 (2017), DOE Supported.

[11] A. Anastasi et al., "The laser control of the muon $g-2$ experiment at Fermilab", JINST **13**, T02009 (2018), DOE Supported.

[12] R. J. Hill, P. Kammel, W. J. Marciano, and A. Sirlin, "Nucleon Axial Radius and Muonic Hydrogen", (2017) {https://doi.org/10.1088/1361-6633/aac190}, DOE Supported.

[13] V. Ganzha, K. Ivshin, P. Kammel, P. Kravchenko, P. Kravtsov, C. Petitjean, V. Trofimov, A. Vasilyev, A. Vorobyov, M. Vznuzdaev, and F. Wauters, "Measurement of trace impurities in ultra pure hydrogen and deuterium at the parts-per-billion level using gas chromatography", Nuclear Instruments and Methods in Physics Research Section A: Accelerators, Spectrometers, Detectors and Associated Equipment **880**, 181 –187 (2018), DOE Supported.

[14] G. Adhikari, P. Adhikari, E. B. de Souza, N. Carlin, S. Choi, W. Choi, M. Djamal, A. Ezeribe, C. Ha, I. Hahn, A. Hubbard, E. Jeon, J. Jo, H. Joo, W. Kang, W. Kang, M. Kauer, B. Kim, H. Kim, H. Kim, K. Kim, M. Kim, N. Kim, S. Kim, Y. Kim, Y. Kim, V. Kudryavtsev, H. Lee, J. Lee, J. Lee, M. Lee, D. Leonard, K. Lim, W. Lynch, R. Maruyama, F. Mouton, S. Olsen, H. Park, H. Park, J. Park, K. Park, W. Pettus, Z. Pierpoint, H. Prihtiadi, S. Ra, F. Rogers, C. Rott, A. Scarff, N. Spooner, W. Thompson, L. Yang, and S. Yong, "Initial Performance of the COSINE-100 Experiment", Eur. Phys. J. **C78**, 107 (2018).

[15] H. Prihtiadi, G. Adhikari, P. Adhikari, E. B. de Souza, N. Carlin, S. Choi, W. Choi, M. Djamal, A. Ezeribe, C. Ha, I. Hahn, A. Hubbard, E. Jeon, J. Jo, H. Joo, W. Kang, W. Kang, M. Kauer, B. Kim, H. Kim, H. Kim, K. Kim, N. Kim, S. Kim, Y. Kim, Y. Kim, V. Kudryavtsev, H. Lee, J. Lee, J. Lee, M. Lee, D. Leonard, K. Lim, W. Lynch, R. Maruyama, F. Mouton, S. Olsen, H. Park, H. Park, J. Park, K. Park, W. Pettus, Z. Pierpoint, S. Ra, F. Rogers, C. Rott, A. Scarff, N. Spooner, W. Thompson, L. Yang, and S. Yong, "Muon detector for the COSINE-100 experiment", JINST **13**, T02007 (2018).

[16] S. Triambak, L. Phuthu, A. García, G. C. Harper, J. N. Orce, D. A. Short, S. P. R. Steininger, A. Diaz Varela, R. Dunlop, D. S. Jamieson, W. A. Richter, G. C. Ball, P. E. Garrett, C. E. Svensson, and C. Wrede, "2_1^+ to 3_1^+ γ width in ^{22}Na and second class currents", Phys. Rev. C **95**, 035501 (2017), DOE Supported.

[17] R. Hong, A. Leredde, Y. Bagdasarova, X. Fléchard, A. García, A. Knecht, P. Müller, O. Naviliat-Cuncic, J. Pedersen, E. Smith, M. Sternberg, D. W. Storm, H. E. Swanson, F. Wauters, and D. Zumwalt, "Charge-state distribution of Li ions from the β decay of laser-trapped ^{6}He atoms", Phys. Rev. A **96**, 053411 (2017), DOE Supported.

[18] K. P. Hickerson, X. Sun, Y. Bagdasarova, D. Bravo-Berguño, L. J. Broussard, M. A.-P. Brown, R. Carr, S. Currie, X. Ding, B. W. Filippone, A. García, P. Geltenbort, J. Hoagland, A. T. Holley, R. Hong, T. M. Ito, A. Knecht, C.-Y. Liu, J. L. Liu, M. Makela, R. R. Mammei, J. W. Martin, D. Melconian, M. P. Mendenhall, S. D. Moore, C. L. Morris, R. W. Pattie, A. Pérez Galván, R. Picker, M. L. Pitt, B. Plaster,

J. C. Ramsey, R. Rios, A. Saunders, S. J. Seestrom, E. I. Sharapov, W. E. Sondheim, E. Tatar, R. B. Vogelaar, B. VornDick, C. Wrede, A. R. Young, and B. A. Zeck, "First direct constraints on Fierz interference in free-neutron β decay", Phys. Rev. C **96**, 042501 (2017), DOE Supported.

[19] M. A.-P. Brown, E. B. Dees, E. Adamek, B. Allgeier, M. Blatnik, T. J. Bowles, L. J. Broussard, R. Carr, S. Clayton, C. Cude-Woods, S. Currie, X. Ding, B. W. Filippone, A. García, P. Geltenbort, S. Hasan, K. P. Hickerson, J. Hoagland, R. Hong, G. E. Hogan, A. T. Holley, T. M. Ito, A. Knecht, C.-Y. Liu, J. Liu, M. Makela, J. W. Martin, D. Melconian, M. P. Mendenhall, S. D. Moore, C. L. Morris, S. Nepal, N. Nouri, R. W. Pattie, A. Pérez Galván, D. G. Phillips, R. Picker, M. L. Pitt, B. Plaster, J. C. Ramsey, R. Rios, D. J. Salvat, A. Saunders, W. Sondheim, S. J. Seestrom, S. Sjue, S. Slutsky, X. Sun, C. Swank, G. Swift, E. Tatar, R. B. Vogelaar, B. VornDick, Z. Wang, J. Wexler, T. Womack, C. Wrede, A. R. Young, and B. A. Zeck, "New result for the neutron β-asymmetry parameter A_0 from UCNA", Phys. Rev. C **97**, 035505 (2018), DOE Supported.

[20] S. Bauer et al., "Reduction of stored-particle background by a magnetic pulse method at the KATRIN experiment", (2018).

[21] C. E. Griggs, M. V. Moody, R. S. Norton, H. J. Paik, and K. Venkateswara, "Sensitive superconducting gravity gradiometer constructed with levitated test masses", Phys. Rev. Applied **8**, 064024 (2017).

[22] G. Ban, G. Bison, K. Bodek, M. Daum, M. Fertl, B. Franke, Z. Grujić, W. Heil, M. Horras, M. Kasprzak, Y. Kermaidic, K. Kirch, H.-C. Koch, S. Komposch, A. Kozela, J. Krempel, B. Lauss, T. Lefort, A. Mtchedlishvili, G. Pignol, F. Piegsa, P. Prashanth, G. Quéméner, M. Rawlik, D. Rebreyend, D. Ries, S. Roccia, D. Rozpedzik, P. Schmidt-Wellenburg, N. Severijns, A. Weis, G. Wyszynski, J. Zejma, and G. Zsigmond, "Demonstration of sensitivity increase in mercury free-spin-precession magnetometers due to laser-based readout for neutron electric dipole moment searches", Nuclear Instruments and Methods in Physics Research Section A: Accelerators, Spectrometers, Detectors and Associated Equipment **896**, 129 –138 (2018).

[23] M. Arenz et al., "First transmission of electrons and ions through the katrin beamline", Journal of Instrumentation **13**, P04020 (2018).

[24] M. Arenz et al., "Calibration of high voltages at the ppm level by the difference of ^{83m}Kr conversion electron lines at the KATRIN experiment", Eur. Phys. J. **C78**, 368 (2018).

[25] M. Arenz et al., "The KATRIN Superconducting Magnets: Overview and First Performance Results", (2018).

[26] P. Adhikari et al., "Background model for the NaI(Tl) crystals in COSINE-100", Eur. Phys. J. **C78**, 490 (2018), DOE Supported.

9.2 Invited talks at conferences

[27] W. C. Pettus, *The liquid argon veto for LEGEND*, invited, Low-Radioactivity Underground Argon Workshop, PNNL, Richland, WA, March 2018, 2018, DOE Supported.

[28] J. Detwiler, *Status of neutrinoless double-beta decay experiments*, invited, Invited talk at the INT Program INT-18-1a on Nuclear ab initio Theories and Neutrino Physics, Seattle, WA (March 2018), 2018, DOE Supported.

[29] J. Detwiler, *Searching for matter creation*, invited, Physics Department Colloquium, Gonzaga University, Spokane, WA (December 2017), 2017, DOE Supported.

[30] J. Detwiler, *First results from the MAJORANA DEMONSTRATOR*, invited, TRIUMF Colloquium, Vancouver, B. C., Canada (November 2017), 2017, DOE Supported.

[31] J. Detwiler, *How a 'jelly doughnut' may explain why the universe exists*, invited, Interview on Sound Effect by Gabriel Spitzer of KNKX (NPR) / 88.5 FM Seattle in (November 2017), 2017, DOE Supported.

[32] J. Detwiler, *Background results from the MAJORANA DEMONSTRATOR*, invited, Conference on Neutrino and Nuclear Physics (CNNP2017), Catania, Italy (October 2017), 2017, DOE Supported.

[33] J. Detwiler, *The quest for neutrinoless double-beta decay*, invited, seminar at the Kavli Institute for the Physics and Mathematics of the Universe, Kashiwa, Japan (August 2017), 2017, DOE Supported.

[34] J. Detwiler, *0nbb searches with 76ge*, invited, invited talk at the INT Program INT-17-2a on Neutrinoless Double-beta Decay, Seattle, WA (June 2017), 2017, DOE Supported.

[35] J. Detwiler, *The MAJORANA DEMONSTRATOR neutrinoless double-beta decay search*, invited, Center for Neutrino Physics Seminar, Virginia Polytechnic Institute and State University, Blacksburg, VA (March 2017), 2017, DOE Supported.

[36] M. Fertl, *Project 8: A frequency-based approach to measure the absolute neutrino mass scale*, Albert Einstein Center Seminar, University of Berne, Switzerland, April 2017.

[37] M. Fertl, *Project 8: A frequency based approach to measure the neutrino mass*, International School of Nuclear Physics, 39th Course Neutrinos in Cosmology, in Astro-, Particle- and Nuclear Physics, Erice-Sicily, Italy, September 2017, DOE Supported.

[38] M. Fertl, *Clocks to weigh Beyond Standard Model Physics: Muon g-2 and the Neutrino Mass Scale*, Univeristy of Washington, Physics colloquium, April 2017, DOE Supported.

[39] M. Fertl, *Peeking beyond the Standard Model: Muon g-2, Neutrino Mass Scale, and Precision β-Decay*, California Institute of technology, Pasadena, CA, HEP seminar, march 2018.

[40] L. Kippenbrock, *Background from the inter-spectrometer Penning trap in the KATRIN experiment*, Pittsburgh, PA: 2017 Fall Meeting of the APS Division of Nuclear Physics, DOE Supported.

[41] L. Kippenbrock, *Status of the KATRIN experiment*, Lake Louise, AB: Lake Louise Winter Institute, DOE Supported.

[42] D. W. Hertzog, F. Symmetries, and P. Physics, *Three 1*, 5-hour lectures at the National Nuclear Physics Summer School in Boulder, Colorado, July, 2017.

[43] D. W. Hertzog, *Report from NSAC: Status since the 2015 Long Range Plan*, IUPAP meeting, (invited presentation), Tokyo, Japan, August 29, 2017.

[44] D. W. Hertzog, *Next-Generation Muon g-2: An indirect, but highly sensitive search for New Physics*, Seminar, Michigan State University, November 8, 2017.

[45] D. W. Hertzog, *Next-Generation Muon g-2: An indirect, but highly sensitive search for New Physics*, Colloquium, Argonne National Laboratory, November 10, 2017.

[46] D. W. Hertzog, *The (anomalous) magnetic moment of the muon*, Symposium on Fundamental Physics in Memory of Sidney Drell, January 12, 2018.

[47] D. W. Hertzog, *Next-Generation Muon g-2: An indirect, but highly sensitive search for New Physics*, Colloquium, Carnegie Mellon University, March 19, 2018.

[48] D. W. Hertzog, *Next-Generation Muon g-2: An indirect, but highly sensitive search for New Physics*, Colloquium and Chertok Lecture, Boston University, April 10, 2018.

[49] G. Rybka, *First results from ADMX g2*, Fermilab Joint Experimental-Theoretical Physics Seminar, April 27, 2018.

[50] G. Rybka, *Project 8*, Lake Louise Winter Institute, Lake Louise, AB, Canada. February, 2018.

[51] G. Rybka, *Dfsz axion dark matter sensitivity with ADMX*, Axions at the crossroads, ECT* Trento, Italy, November 20-24, 2017.

[52] G. Rybka, *The ADMX search for dark matter axions*, University of Victoria Seminar, Victoria, BC, October 18, 2017.

[53] G. Rybka, *Comments from the US cosmic visions workshop*, 13th Patras Workshop on Axions Wimps and Wisps, Thessaloniki, Greece, May 19, 2017.

[54] G. Rybka, *Searching for dark matter with ADMX g2 and beyond*, DOE Cosmic Visions Workshop, Maryland, Mar. 2017.

[55] W. C. Pettus, *Overview of project 8 and progress towards tritium operation*, 15th International Conference on Topics in Astroparticle and Underground Physics (TAUP 2017) Sudbury, Ontario, Canada, July 24-28, 2017, DOE Supported.

[56] R. Robertson, *Progress toward a direct measurement of neutrino mass*, Invited talk, Baksan 50th Anniversary Symposium, Baksan, Russia, 2017.

[57] R. Robertson, *Remarks*, Commencement address, McMaster University, Hamilton ON, 2017.

[58] R. Robertson, *Progress toward measuring the mass of the neutrino*, Colloquium, University of Illinois, Champaign-Urbana IL, 2017.

[59] Y.-T. Lin, *Trims: validating t_2 molecular effects for neutrino mass experiments*, APS April Meeting 2017, Washington, DC, 2017.

[60] Y.-T. Lin, *Trims: validating t_2 molecular effects for neutrino mass experiments*, DNP Meeting 2017, Pittsburgh, PA, 2017.

9.3 Abstracts and contributed talks

[61] R. Robertson, *Project 8: a new approach to measuring neutrino mass*, Contributed talk, APS-DNP meeting, Pittsburgh, PA, 2017.

9.4 Reports, white papers and proceedings

[62] COHERENT Collaboration, D. Akimov, J. B. Albert, P. An, C. Awe, P. S. Barbeau, B. Becker, V. Belov, M. A. Blackston, A. Bolozdynya, A. Brown, A. Burenkov, B. Cabrera-Palmer, M. Cervantes, J. I. Collar, R. J. Cooper, R. L. Cooper, C. Cuesta, J. Daughhetee, D. J. Dean, J. del Valle Coello M. Detwiler and, M. D'Onofrio, A. Eberhardt, Y. Efremenko, S. R. Elliott, A. Etenko, L. Fabris, M. Febbraro, N. Fields, W. Fox, Z. Fu, A. Galindo-Uribarri, M. P. Green, M. Hai, M. R. Heath, S. Hedges, D. Hornback, T. W. Hossbach, E. B. Iverson, M. Kaemingk, L. J. Kaufman, S. R. Klein, A. Khromov, S. Ki, A. Konovalov, A. Kovalenko, M. Kremer, A. Kumpan, C. Leadbetter, L. Li, W. Lu, K. Mann, D. M. Markoff, Y. Melikyan, K. Miller, H. Moreno, P. E. Mueller, P. Naumov, J. Newby, J. L. Orrell, C. T. Overman, D. S. Parno, S. Penttila, G. Perumpilly, D. C. Radford, R. Rapp, H. Ray, J. Raybern, D. Reyna, G. C. Rich, D. Rimal, D. Rudik, D. J. Salvat, K. Scholberg, B. Scholz, G. Sinev, W. M. Snow, V. Sosnovtsev, A. Shakirov, S. Suchyta, B. Suh, R. Tayloe, R. T. Thornton, I. Tolstukhin, J. Vanderwerp, R. L. Varner, C. J. Virtue, Z. Wan, J. Yoo, C.-H. Yu, A. Zawada, A. Zderic, and J. Zettlemoyer, "COHERENT Collaboration data release from the first observation of coherent elastic neutrino-nucleus scattering", ArXiv e-prints (2018), DOE Supported.

[63] J. Carlson, M. J. Savage, J. A. Detwiler, et al., "Nuclear Physics Exascale Requirements Review Report", US DOE NP/ASCR Meeting Report (2017), DOE Supported.

[64] W. C. Pettus, "Overview of Project 8 and Progress Towards Tritium Operation", in 15th International Conference on Topics in Astroparticle and Underground Physics (TAUP 2017) Sudbury, Ontario, Canada, July 24-28, 2017 (2017), DOE Supported.

[65] A Ashtari Esfahani et al., "Project 8 detector upgrades for a tritium beta decay spectrum using cyclotron radiation", in Proceedings, 27th International Conference on Neutrino Physics and Astrophysics (Neutrino 2016): London, United Kingdom, July 4-9, 2016 (2017).

[66] A. Ashtari Esfahani et al., "Results from the Project 8 phase-1 cyclotron radiation emission spectroscopy detector", in Proceedings, 27th International Conference on Neutrino Physics and Astrophysics (Neutrino 2016): London, United Kingdom, July 4-9, 2016, Vol. 888, 1 (2017), p. 012074.

[67] A. Ashtari Esfahani et al., "Project 8 Phase III Design Concept", in Proceedings, 27th International Conference on Neutrino Physics and Astrophysics (Neutrino 2016): London, United Kingdom, July 4-9, 2016, Vol. 888, 1 (2017), p. 012230.

9.5 Ph.D. degrees granted

[68] B. M. Graner, "Reduced limit on the permanent electric dipole moment of 199 hg", http://hdl.handle.net/1773/40676, PhD thesis (University of Washington, Department of Physics, Aug. 2017).

[69] J. Gruszko, "Surface Alpha Interactions in P-Type Point-Contact HPGe Detectors: Maximizing Sensitivity of ^{76}Ge Neutrinoless Double-Beta Decay Searches", http://hdl.handle.net/1773/40682, PhD thesis (University of Washington, Department of Physics, Aug. 2017).

[70] E. Lentz, "Improving axion signal models through n-body simulations", http://hdl.handle.net/1773/40961, PhD thesis (University of Washington, Department of Physics, Oct. 2017).

[71] E. Martin, "Electron detection systems for katrin detector and spectrometer section", http://hdl.handle.net/1773/40285, PhD thesis (University of Washington, Department of Physics, May 2017).

[72] M. W. Smith, "Developing the precision magnetic field for the e989 muon g–2 experiment", http://hdl.handle.net/1773/40681, PhD thesis (University of Washington, Department of Physics, Aug. 2017).

[73] M. D. Turner, "Development of new technologies for precision torsion-balance experiments", http://hdl.handle.net/1773/40958, PhD thesis (University of Washington, Department of Physics, Nov. 2017).

Back row: Peter Doe, Jim Elms, Sanshiro Enomoto, Seth Kimes, Nick Ruof, Jason Detwiler, Gray Rybka, John Lee, Kim Siang Khaw, Ethan Muldoon, Daniel Salvat, Alvaro Chavarria, Daniel Garratt, Derek Storm, Tim Van Wechel, David Peterson, Brynn MacCoy, Matthew Kallander, Ian Guinn, Roland Farrell

Second-to-back row: Minjung Sung, Tom Burritt, Zachary Wuthrich, Aaron Fienberg, Walter Pettus, Sean Fell, Gary Holman, Jared Johnson, Pitam Mitra, Raphael Cervantes, Alexandru Hostiuc, Jason Hempstead, Eric Machado, Celia Chor, Nerissa Pineda, Christine Claessens, Joben Pedersen

Sitting (chairs): Hannah Binney, Yelena Bagdasarova, Eric Adelberger, John Cramer, Eric Swanson, Doug Will, David Hertzog, Hamish Robertson, Charlie Hagedorn, Brent Graner, Eric Smith, Gerry Garvey, Nick Force

Sitting (ground): Nick Du, Megan Ivory, Erik Shaw, Krishna Venkateswara, Micah Buuck, Michael Huehn, Rachel Ryan, Peter Kammel, Nathan Froemming, Alejandro Garcia, Elise Novitski, Luke Kippenbrock

www.ingramcontent.com/pod-product-compliance
Lightning Source LLC
Chambersburg PA
CBHW051146220526
45473CB00003B/672